PRACTICAL
PROBLEMS *in*
MATHEMATICS
for WELDERS

Delmar's *PRACTICAL PROBLEMS in MATHEMATICS* Series

- *Practical Problems in Mathematics for Automotive Technicians, 4e*
  George Moore
  Order # 0-8273-4622-0

- *Practical Problems in Mathematics for Carpenters, 5e*
  Harry C. Huth
  Order # 0-8273-4579-8

- *Practical Problems in Mathematics for Electricians, 5e*
  Herman and Garrard
  Order 0-8273-6708-2

- *Practical Problems in Mathematics for Electronic Technicians, 3e*
  Herman and Sullivan
  Order # 0-8273-6761-9

- *Practical Problems in Mathematics for Graphic Artists*
  Vermeersch and Southwick
  Order # 0-8273-2100-7

- *Practical Problems in Mathematics for Heating and Cooling Technicians, 2e*
  Russell B. DeVore
  Order # 0-8273-4062-1

- *Practical Problems in Mathematics for Manufacturing, 4e*
  Dennis D. Davis
  Order # 0-8273-6710-4

- *Practical Problems in Mathematics for Masons, 2e*
  John E. Ball
  Order # 0-8273-1283-0

- *Practical Problems in Mathematics for Mechanical Drafting*
  John C. Larkin
  Order # 0-8273-1670-4

- *Practical Problems in Mathematics for Welders, 4e*
  Schell and Matlock
  Order # 0-8273-6706-6

Related Titles

- *Mathematics for the Automotive Trade, 2e*
  Peterson and DeKryger
  Order # 0-8273-3554-7

- *Mathematics for Electricity and Electronics*
  Dr. Arthur Kramer
  Order 0-8273-5804-0

PRACTICAL PROBLEMS in MATHEMATICS
for WELDERS
4th edition

Frank R. Schell

Bill J. Matlock

Delmar Publishers

I(T)P An International Thomson Publishing Company

Albany • Bonn • Boston • Cincinnati • Detroit • London • Madrid • Melbourne
Mexico City • New York • Pacific Grove • Paris • San Francisco • Singapore • Tokyo
Toronto • Washington

NOV 0 1 1999

NOTICE TO THE READER

Cover Credit: Dartmouth Publishing Inc.

Delmar Staff:
Publisher: Michael A. McDermott
Editor: Mary Clyne
Art & Design Coordinator: Nicole Reamer
Production Manager: Larry Main

COPYRIGHT © 1996
By Delmar Publishers Inc.
a division of International Thomson Publishing Inc.
The ITP logo is a trademark under license.

Printed in the United States of America

For more information, contact:
Delmar Publishers
3 Columbia Circle, Box 15015
Albany, New York 12212-5015

International Thomson Editores
Campos Eliseos 385, Piso 7
Col Polanco
11560 Mexico D F Mexico

International Thomson Publishing Europe
Berkshire House 168 - 173
High Holborn
London WC1V 7AA
England

International Thomson Publishing GmbH
Königswinterer Strasse 418
53227 Bonn
Germany

Thomas Nelson Australia
102 Dodds Street
South Melbourne, 3205
Victoria, Australia

International Thomson Publishing Asia
221 Henderson Road
#05 - 10 Henderson Building
Sinapore 0315

Nelson Canada
1120 Birchmount Road
Scarborough, Ontario
Canada M1K 5G4

International Thomson Publishing - Japan
Hirakawacho Kyowa Building, 3F
2-2-1 Hirakawacho
Chiyoda-ku, Tokyo 102
Japan

Library of Congress Cataloging-in-Publication Data

Schell, Frank R.
    Practical problems in mathematics for welders /Frank R. Schell,
Bill J. Matlock. -- 4th ed.
        p.   cm.
    ISBN: 0-8273-6706-6
    1. Welding--Mathematics.   I. Matlock, Bill J., 1938-
    II. Title.
TS227.2.S37   1995
671.5'2'0151--dc20   95-12431
                                                         CIP

Contents

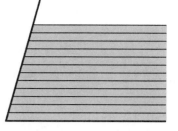

SECTION 5 DIRECT MEASURE

SECTION 6 ANGULAR MEASURE

SECTION 7 COMPUTED MEASURE

SECTION 8 PIECES AND LENGTHS

Preface

It is of great advantage to be able to apply mathematical skills to welding problems that arise during the fabrication of materials into welded structures. Calculation for all problems encountered may vary, but essentially the competent worker follows a regular pattern of procedure in all work. The problems presented are essentially the same. In analyzing the job, there are certain aspects that can follow through any job.

1. Decide on the outcome to be achieved.

2. Determine the amount of accuracy necessary for the job; that is, does it need to be as near perfect, mathematically, as possible, or is there room for some variation? Here the blueprint should give instructions for tolerances—the plus or minus amount allowed for fit-up on the completed job. It is necessary for the welder to be aware of any errors in calculations, and all calculations should be checked a second time.

Practical Problems in Mathematics for Welders, Fourth Edition, was prepared by welders who have had many years of experience in teaching vocational welding students. All of the relevant and easily understandable problems in this workbook relate to actual shop situations. This material is designed to provide the student with valuable and necessary practice in solving typical welding problems.

Practical problems in Mathematics for Welders, Fourth Edition, is designed for use by a wide range of students. The workbook is suitable for any student from the junior high school level through high school and the two-year college level. This kind of book has many benefits for the instructor and for the student. For the student, the workbooks offer a step-by-step approach to the mastery of essential skills in mathematics.

At the outset of each unit, an introductory section reviews mathematics principles and procedures to be used in solving the problems.

Each workbook is complemented by an Instructor's Guide, which includes answers to every problem in the Workbook and solutions to many of the more complicated problems. Two Achievement Reviews are provided in the Instructor's Guide to provide effective means of measuring the student's progress.

Appendix material provides reference material for the welding student's use.

Whole Numbers

Unit 1   ADDITION OF WHOLE NUMBERS

BASIC PRINCIPLES

The addition of whole numbers is a procedure necessary for all welders to use. When adding whole numbers, the following steps apply:

Whole numbers 9 and below are placed one beneath the other. The plus sign (+) is used.

Example:
```
    3
    6
  + 9
   18
```

When numbers over 9 are added, the numbers are arranged in columns, adding the numbers from right to left:

Example: 3 + 6+ 9 + 14 + 214 =

```
    3
    6
    9
   14
 + 214
```

If the answer in the first column is a number of greater value than 9, the second digit is carried to the top of the second column and added to that column. The process continues until all columns have been added.

Example:

```
    3                    2              2
    3          3         3             3
    6          6         6             6
    9          9         9             9
   14          4        14            14
 + 214         4      + 214         + 214
              26         6           246
```

PRACTICAL PROBLEMS

1.   An inventory of the steel rack of a welding shop shows: angle, 98 feet; channel, 193 feet; I beam, 84 feet; 1-inch square tubing, 1,127 feet. Find, in feet, the total amount of steel in the inventory.

_____

2.   Layout work for a welded rectangular steel pipe is shown. Determine the total number of inches of steel in the length of the layout.

_____

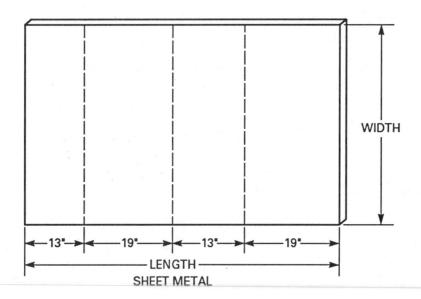

3.   A welded steel framework consists of: plate steel, 1,098 pounds; bolt stock, 98 pounds; key stock, 13 pounds; channel, 10,114 pounds. Find the total number of pounds of steel in the framework.

_____

Problem: 1,098 pounds
                     98 pounds
                     13 pounds
          + 10,114 pounds

Note: To identify steel shapes, refer to the Glossary.

One method of checking addition for a correct answer is the use of the "cast out" formula. Refer to the problem on page 1.

Add the numbers in each *horizontal* row. Reduce the answers to a single number. Keep adding until the numbers are all single digits.

$$3 \qquad\qquad\qquad = 3$$

$$6 \qquad\qquad\qquad = 6$$

$$9 \qquad\qquad\qquad = 9$$

$$14 \quad (4+1) = 5 \qquad = 5$$

$$214 \quad (4 + 1 + 2) = 7 \qquad = \underline{7}$$

$$246 \qquad\qquad\qquad\qquad 30 = ③ \text{(single digit)}$$

$$2 + 4 + 6 = 12$$

$$(12) = 1 + 2 = ③$$

If the single digits are the same, the original answer is correct.

Unit 2   SUBTRACTION OF WHOLE NUMBERS

BASIC PRINCIPLES

For practical welding it is often necessary to subtract one measurement from another. The minus sign (-) is used to indicate subtraction. The number being subtracted is called the *subtrahend.* The number being subtracted from is called the *minuend.* The number that is left is called the *difference.*

The subtrahend is placed below the minuend. The numbers are lined up from the right, as in addition.

Example:

$$
\begin{array}{r}
3 \\
-\ 2 \\
\hline
1 \\
\end{array}
\quad
\begin{array}{l}
\text{(minuend)} \\
\text{(subtrahend)} \\
\text{(difference)} \\
\end{array}
$$

Beginning on the right-hand side, if the bottom number is greater in value than the number above it, it will be necessary to borrow one *unit* from the next column in the minuend.

$$
\begin{array}{ccc}
 & & \text{Becomes 10} \\
 & 7\!\downarrow & 7 \\
180 & 180 & 180 \\
-\ 29 & =\ -\ 29 & -\ 29 \\
 & & \hline \\
 & & 151 \\
\end{array}
$$

(By bringing one unit from the second column it reduces the 8 to a 7, and adds *10* to the first column, thus making the subtraction possible.)

To prove the answer correct in problems of subtraction, add the *difference* to the original *subtrahend.* The answer should equal the *minuend* of the original problem.

$$
\begin{array}{r}
151 \\
+\ 29 \\
\hline
180 \\
\end{array}
\quad
\begin{array}{l}
\text{(difference)} \\
\text{(subtrahend)} \\
\text{(original minuend)} \\
\end{array}
$$

PRACTICAL PROBLEMS

1. A welder is required to flame cut a piece of angle iron as shown in the illustration. After the cut piece is removed, how much angle iron remains from the original piece, in inches? Allow ¼ inch waste for the cut.

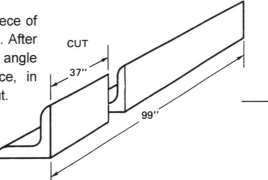

CUT

37"

99"

STEEL ANGLE

_____

2. The inventory of a scrap pile is as follows: channel iron, 89 inches; plate steel, 34 inches; key stock, 116 inches; pipe, 65 inches; flat strap iron, 18 inches. A pipe support is welded using the following material: channel iron, 61 inches; plate steel, 9 inches; key stock, 39 inches; pipe, 38 inches; flat strap iron, 9 inches. What is the balance of material remaining in the scrap pile, in inches?

    Channel iron   a. _____

    Plate steel   b. _____

    Key stock   c. _____

    Pipe   d. _____

    Flat strap iron   e. _____

3. A length of pipe is cut as shown in the illustration. After the two cut pieces are removed, how long is the remaining length of pipe, in inches? Allow ⅛ inch waste for each cut.   _____

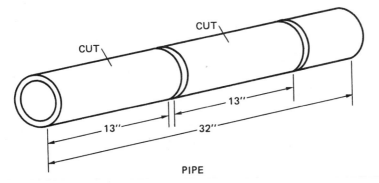

CUT

CUT

13"

13"

32"

PIPE

4.   A stock room has 18,903 pounds of plate steel. A steel storage tank is welded from 1,366 pounds of plate. How many pounds of plate remains in stock?   _____

Unit 3   MULTIPLICATION OF WHOLE NUMBERS

BASIC PRINCIPLES AND PROCEDURES

Multiplication is the same as addition except that it is a shorter method.

a. $4 + 4 + 4 = 12$ b. $4 \times 3 = 12$

Each part of the problem has a name. The top (or first) number – the one being multiplied – is called the *multiplicand*. The lower (or second) number – the one that tells how many times the first number is to be "added" to itself – is called the *multiplier*. The result is called the *product*.

To simplify addition, use the number to multiply itself by the number of times needed. The symbol x (times) is used to indicate multiplication.

c.   If the numbers are larger than 9, the remainders must be carried to the next column.

Thus:
$$
\begin{array}{r}
9 \\
\times\ 9 \\
\hline
81
\end{array}
\quad
\begin{array}{l}
\text{(Multiplicand)} \\
\text{(Multiplier)} \\
\text{(Product)}
\end{array}
$$

But:
$$
\begin{array}{r}
1 \downarrow \\
95 \\
\times\ 3 \\
\hline
285
\end{array}
$$

Explanation: $3 \times 5 = 15$. Place the 5 in the answer, carry the 1 to the next column. $3 \times 9 = 27 +$ the 1 carried over $= 28$. The answer is 285.

Proof:
$$
\begin{array}{r}
95 \\
95 \\
+\ 95 \\
\hline
285
\end{array}
$$

PRACTICAL PROBLEMS

1. A welded support is illustrated.

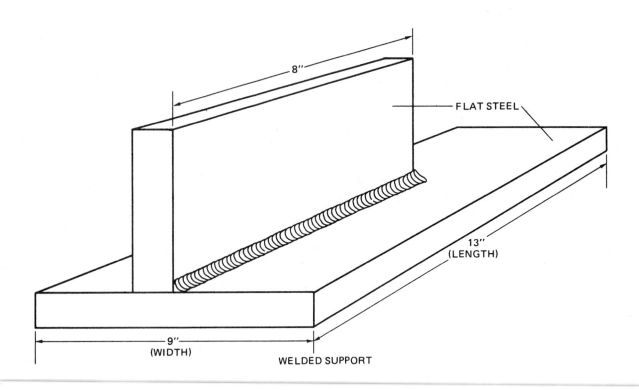

FLAT STEEL

8"

13"
(LENGTH)

9"
(WIDTH)

WELDED SUPPORT

a. Find, in inches, the total length of weld required for 660 supports. a. _____

b. The support plate is 13 inches long and 9 inches wide. How much
    9-inch-wide plate stock, in inches, is used for the completed order? b. _____

c. Each support weighs 4 pounds. What is the weight in pounds of the
    total order of 660 supports? c. _____

2. A welded tank support requires 14 pieces of wide-flange beam to be cut.
    Each piece of beam is 33 inches long. What is the total number of
    inches of beam used? Disregard waste caused by the width of the cut. _____

3. A job requires 1,098 pieces of bar stock, each 9 inches long. What is the
    total length of bar stock required, in inches? _____

4. A welder tack welds 215 linear feet of steel support columns per hour.
    How many feet of supports are completed in an eight-hour shift? _____

Unit 4   DIVISION OF WHOLE NUMBERS

BASIC PRINCIPLES

The signs used to indicate division are $\overline{)}$, ——, and ÷. The number to be divided is called the *dividend.* The number used to divide the dividend is called the *divisor.* The number resulting from the division is called the *quotient.*

With the number 10 as the divisor, and the number 110 as the dividend, the problem appears as follows:

$$10\ \overline{)110}$$

How many times is 10 contained in 110? First find out how many tens are contained in 11. The answer (1) is placed over the ones place of 11 above the division box.

$$
\begin{array}{r}
1 \\
10\ \overline{)110}
\end{array}
$$

1 times 10 equals 10. The ten is placed below the 11 and subtracted and the remainder (1) is written below.

$$
\begin{array}{r}
1 \\
10\ \overline{)110} \\
-\,10 \\
\hline
1
\end{array}
$$

The zero (0) remaining in the dividend is dropped down next to the one (1).

$$
\begin{array}{r}
11 \\
10\ \overline{)110} \\
-\,10 \\
\hline
10
\end{array}
$$

Since ten (10) is contained in ten (10) one time, the one (1) is placed above the divisor box as part of the quotient.

$$
\begin{array}{r}
11 \\
10\ \overline{)110} \\
-\,10 \\
\hline
10
\end{array}
$$

One times 10 equals 10. Subtracted, the remainder is zero (0); therefore, 110 divided by 10 gives a quotient of 11.

$$
\begin{array}{r}
11 \\
10\ \overline{)110} \\
-10 \\
\hline
10 \\
-10 \\
\hline
0
\end{array}
$$

To prove the answer, multiply the divisor by the quotient:

Divisor: 10

Quotient:
$$
\begin{array}{r}
\times 11 \\
\hline
10 \\
+10 \\
\hline
110
\end{array}
$$

PRACTICAL PROBLEMS

1.   New steel is delivered in 21-foot lengths. A section must be sheared into 3-foot-long pieces. How many pieces can be obtained from each length?   _____

2.   A piece of ⅜-inch x 5-inch flat stock has 12 equally spaced holes drilled in it. What is the distance, in inches, between the centers of the holes?   _____

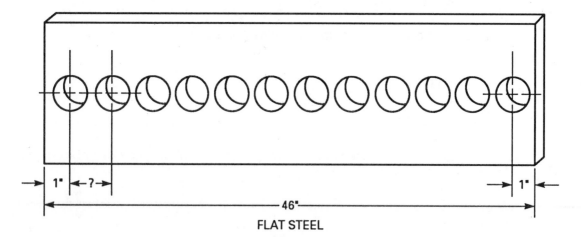

FLAT STEEL

3. A steel support has five holes drilled at equal distances. Find, in inches, the center-to-center distance between the holes. _____

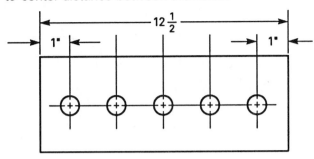

FLAT STEEL

4. How many shear pins, each 8 inches long, can be cut from the round bar? Disregard waste caused by the width of the cuts. _____

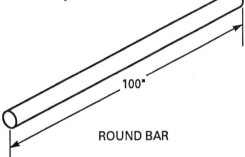

100"

ROUND BAR

5. a. Fifteen water tanks are constructed in a welding shop. All tanks hold a total of 13,500 gallons. How many gallons will each tank hold?   a. _____

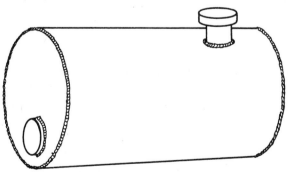

HOT-ROLLED SHEET STEEL

b. The 15 tanks have a total weight of 5,745 pounds. What is the weight of each tank?   b. _____

c. The total contract price for the 15 tanks is $15,540. What is the cost of each tank?   c. _____

Common Fractions

Unit 5   INTRODUCTION TO THE STEEL TAPE AND COMMON FRACTIONS

BASIC PRINCIPLES

The steel tape is used to measure various lengths of steel in inches and fractional parts of the inch. The accurate measuring of structural steel is an important part of the job of the welder.

The steel tape is divided into inches and *fractional* parts of an inch. The most commonly used fractional parts are $1/64$ inch, $1/32$ inch, $1/16$ inch, $1/8$ inch, $1/4$ inch, and $1/2$ inch. (When used, the sign (″) represents inches; the sign (′) represents one foot, or 12 inches.) Each division represents $1/2$ of the next higher division; thus, $1/64$ is one-half of $1/32$; $1/32$ is one-half of $1/16$; $1/16$ is one-half of $1/8$; $1/8$ is one-half of $1/4$ (one-quarter); and $1/4$ is one-half of $1/2$.

If the whole piece is divided into four parts, each part will represent one-quarter ($1/4$); the whole piece being composed of $4/4$, or 1.

A fraction is divided into two parts. The *numerator,* which is above the line (or to the left of the slash), and the *denominator,* which is below the line (or to the right of the slash).

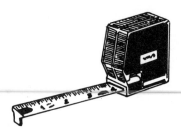

DEFINITION OF A FRACTION

A fraction expresses one or more of the equal parts into which the whole is divided, and it is always written in the form of an indicated division. For example, $3/8$ of an inch means that the inch is divided into 8 equal parts and 3 of these parts are indicated. To express the fact that the fraction $3/8$ indicates division, we can write it thus: $3 \div 8$.

In the fraction $3/8$, the number of equal parts taken (3) is called the numerator of the fraction. The number of equal parts into which the whole is divided (8) is called the denominator of the fraction. In other words, *the denominator is the divisor,* and *the numerator is the dividend.*

NUMERATOR	3	THREE PARTS	DIVIDEND
DENOMINATOR	8	EIGHT PARTS (WHOLE)	DIVISOR

Signs of division are ——, /, and ÷.

The terms of a fraction are the numerator and the denominator, and a common fraction is a fraction in which both terms are used.

COMMON FRACTION $\qquad \dfrac{3}{8} \qquad$ NUMERATOR
DENOMINATOR

If the numerator 3/ of a fraction is less than the denominator /8, the fraction is called a proper fraction. The following are proper fractions:

$$\frac{3}{4}, \frac{3}{5}, \frac{9}{16}, \frac{7}{8}, \frac{2}{3}, \text{ and so forth.}$$

If the numerator 4/ is *equal* to or *greater* than the denominator/4, the fraction is called an improper fraction. The following are improper fractions:

$$\frac{63}{2}, \frac{68}{10}, \frac{25}{3}, \frac{4}{4}, \frac{6}{2},$$

and so forth.

A mixed number is one that is made up of a *whole* number and a *fraction* of a *whole* number. For example, 5 5⁄16; 5 is the whole number and 5⁄16 is the fraction of a whole number.

WHOLE–5 $\qquad \dfrac{5}{16} \qquad$ – FRACTION

REDUCTION OF COMMON FRACTIONS AND MIXED NUMBERS

Rule: The value of a fraction is not changed by *multiplying* the numerator and the denominator by the same quantity. The value of a fraction is not changed by *dividing* the numerator and the denominator of a fraction by the same quantity. The process of reducing a fraction is sometimes called cancellation.

Example: $\qquad \dfrac{2}{5} \times \dfrac{3}{3} = \dfrac{6}{15}$

This answer can be reduced to its lowest terms by dividing the numerator and denominator by the same number.

Thus:

$$\frac{\overset{2}{\cancel{6}}}{\underset{5}{\cancel{15}}} \div \frac{\overset{1}{\cancel{3}}}{\underset{1}{\cancel{3}}} = \frac{2}{5}$$

When both the numerator and the denominator cannot be divided evenly by the same quantity, the fraction is at its lowest terms.

Example: $\dfrac{2}{5} \div \dfrac{2}{2} = \dfrac{1}{?}$

The denominator cannot be evenly divided by two, so the fraction is at its lowest terms.

An improper fraction can be reduced by dividing the denominator into the numerator.

$$\frac{5}{3} \;=\; 3\overline{)5} \;\;= 3\overline{)\underset{\dfrac{3}{2}}{\overset{1}{5}}} \;\;=\; 1\frac{2}{3}$$

Dividing 3 into 5 leaves a remainder of 2.

REDUCTION OF MIXED NUMBERS

Rule: A mixed number such as 1 ³⁄₈ can be changed to an improper fraction (11⁄8) by multiplying the whole number by the denominator and adding the numerator.

Examples: $1\dfrac{3}{8} \;=\; \dfrac{1 \times 8 + 3}{8} \;=\; \dfrac{11}{8}$

$3\dfrac{3}{4} \;=\; \dfrac{3 \times 4 + 3}{4} \;=\; \dfrac{15}{4}$

PRACTICAL PROBLEMS

1. This piece of steel angle is cut into 16 equal parts. Express each as a
    fractional part of the whole.

Note: Each part is $\frac{1}{16}$th of the whole.

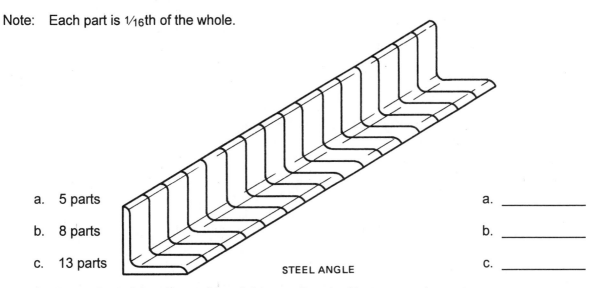

STEEL ANGLE

a.   5 parts   a. _____

b.   8 parts   b. _____

c.   13 parts   c. _____

2. A piece of steel bar is cut into eight equal parts. Express each as a
    fractional part of the whole.

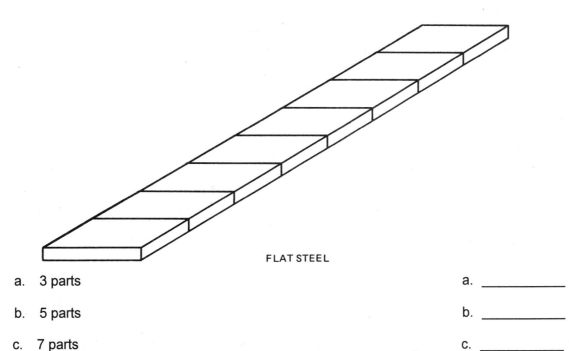

FLAT STEEL

a.   3 parts   a. _____

b.   5 parts   b. _____

c.   7 parts   c. _____

3. What is distance A as shown in the figure? Read the distance from the
    start of the steel tape in the figure to the letter.

A _____

4. What are the distances from the start of the steel tape in the figure to the
    letters B, C, and D?

B _____

C _____

D _____

5. What are the distances from the start of the steel tape in the figure to the
    letters E, F, G, and H?

E _____

F _____

G _____

H _____

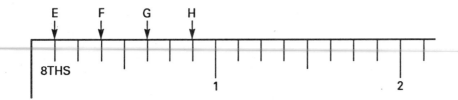

6. Write the distances from the start of the steel tape in the figure to the
    letters I, J, K, L, and M.

I _____

J _____

K _____

L _____

M _____

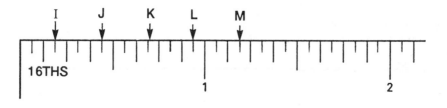

Unit 6   ADDITION OF COMMON FRACTIONS

BASIC PRINCIPLES

In order to find out how much material is needed for a particular job, the addition of common fractions is necessary. Addition, for which the symbol "plus" (+) is used, is the process of finding the total of two or more numbers or fractional parts of numbers. Fractions cannot be added if their denominators are unlike ($3/8$ + $1/16$). Therefore, it is necessary to change all the denominators to the same quantity.

When adding or subtracting fractions, the *least common denominator* or LCD must be found, because to add or subtract fractions they must have the same denominator. To add $1/8''$ to $1/4''$ find the smallest number that both denominators will divide into.

Since 8 is the LCD, the fraction $1/8$ does not change. To find the numerator for the new fraction we are changing from $1/4$, divide the denominator 4 (the old denominator) into the new denominator 8. The number 4 divides evenly into the number 8 two times. Then multiply that number (2) by the old numerator (1). The number 2 multiplied by the number 1 equals 2. So, 2 is the new numerator. Proceed with addition in the usual way, remembering that the denominators are not added, just brought down; it is the numerators that are added.

The product of the addition:

$$\frac{1}{8} = \frac{1}{8}$$
$$+\ \frac{1}{4} = +\ \frac{2}{8}$$
$$\frac{3}{8}$$

Also, $1/16$ + $1/8$ + $1/4$ cannot be added in their present forms. These fractions must all be changed to have the same denominator before adding is possible.

(Hint: When all the fractions to be added, or subtracted, are shop fractions, the LCD will *always* be the largest denominator of the problem.)

Example: Add $1/16'' + 1/8'' + 1/4''$.

$$
\begin{array}{ccc}
\dfrac{1}{16} & = & \dfrac{1}{16} \\[2mm]
\dfrac{1}{8} & = & \dfrac{2}{16} \\[2mm]
+\ \dfrac{1}{4} & = & +\ \dfrac{4}{16} \\[2mm]
& & \dfrac{7}{16}
\end{array}
$$

To add mixed numbers, add the whole numbers and fractions separately; then combine the sums.

Problem: $7\ 5/8 + 4\ 1/2 + 3\ 5/16$

Add the whole numbers.

$$
\begin{array}{r}
7 \\
4 \\
\underline{3} \\
14
\end{array}
$$

Add the fractions after finding the LCD.

$$
\begin{array}{ccccc}
\dfrac{5}{8} & = & \dfrac{5\ \times\ 2}{8\ \times\ 2} & = & \dfrac{10}{16} \\[3mm]
\dfrac{1}{2} & = & \dfrac{1\ \times\ 8}{2\ \times\ 8} & = & \dfrac{8}{16} \\[3mm]
+\ \dfrac{5}{16} & = & \dfrac{5\ \times\ 1}{16\ \times\ 1} & = & +\ \dfrac{5}{16} \\[3mm]
& & & & \dfrac{23}{16}
\end{array}
$$

Reduce the improper fraction to its lowest terms by dividing the denominator into the numerator.

$$
\begin{array}{r}
1 \\
16\ \overline{)\ 23} \\
\underline{16} \\
7
\end{array}
\quad = \quad 1\ 7/16
$$

Add the mixed number that results to the sum of the whole numbers.

$$14 + 1\ 7/16 = 15\ 7/16$$

PRACTICAL PROBLEMS

1. Find the total combined length of these three pieces of steel angle. _____

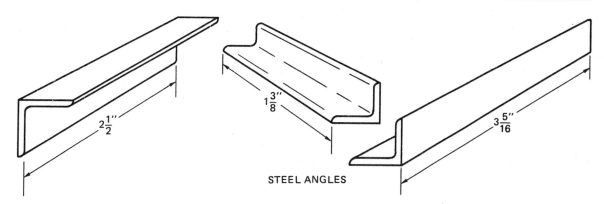

STEEL ANGLES

2. What is the total combined thickness of these four pieces of bar steel? _____

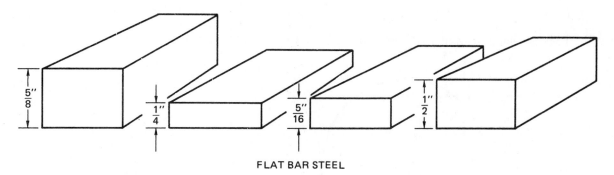

FLAT BAR STEEL

3. Find the total combined width of these four pieces of steel plate. _____

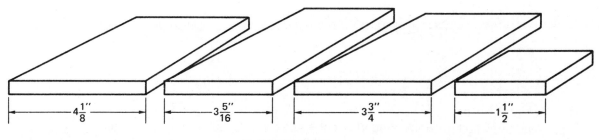

FLAT BAR STEEL

4.   Four holes are drilled in this piece of flat stock. What is the total distance
     between the centers of the holes?   _____

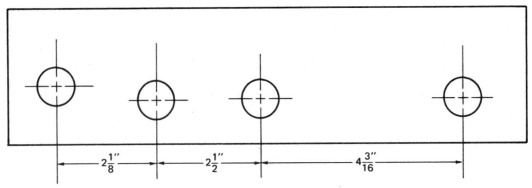

FLAT STEEL

5.   Two circular pieces of steel must be cut from a sheet of ½" plate.

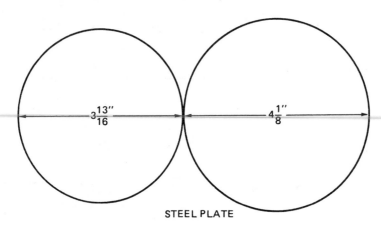

STEEL PLATE

a. Find the length of stock needed to cut these two circles.   a. _____

b. How wide must the stock be?   b. _____

Allow ⅛" waste for each cut. The waste is caused by the width of the
kerf, which is a result of the oxy-fuel cutting process.

6.   To make an off-center bracket, four pieces of ¾″ round stock are welded together. What is the total length of ¾″ material used in the construction?   _____

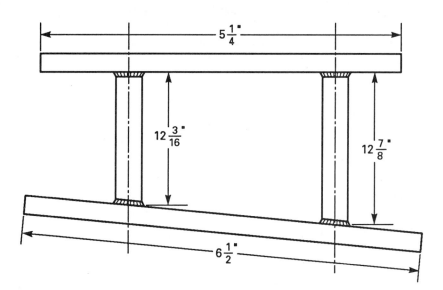

7.   To make shims for leveling a shear, three pieces of material are welded together. What is the total thickness of the welded material, in inches?   _____

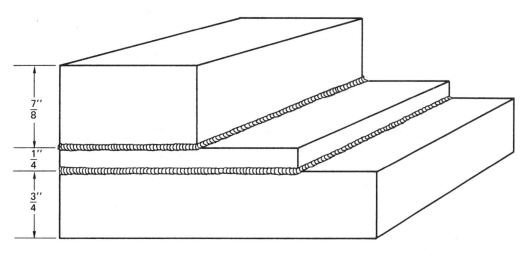

WELDED MATERIAL

8. To make use of some scrap, four pieces of 1½" cold-rolled bar are welded together. What is the total length of the completed weldment? _____

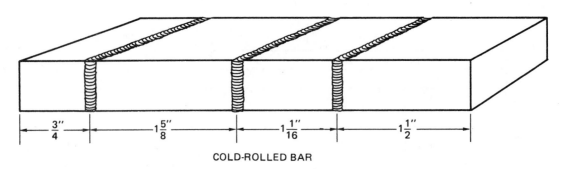

$\frac{3''}{4}$ $1\frac{5''}{8}$ $1\frac{1''}{16}$ $1\frac{1''}{2}$

COLD-ROLLED BAR

9. A motor slide block is welded. Three pieces of bar stock are welded to a plate. What is the total width of the bar stock used, in inches, before drilling bolt holes?

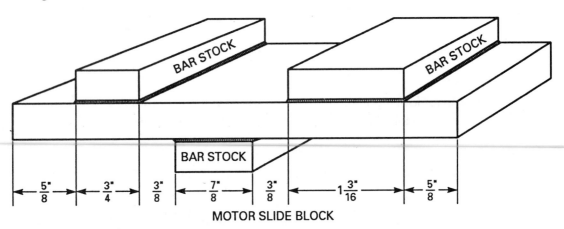

BAR STOCK   BAR STOCK

BAR STOCK

$\frac{5"}{8}$ $\frac{3"}{4}$ $\frac{3"}{8}$ $\frac{7"}{8}$ $\frac{3"}{8}$ $1\frac{3"}{16}$ $\frac{5"}{8}$

MOTOR SLIDE BLOCK

a. **Problem:** Width of the bar stock.

$$\begin{array}{r} \frac{3}{4}'' \\ LCD = 16 \qquad \frac{7}{8}'' \\ +\ 13\!/\!16'' \\ \hline \end{array}$$

a. _____

b. **Problem:** What is the length of the plate used for the weldment?   b. _____

Unit 7 SUBTRACTION OF COMMON FRACTIONS

BASIC PRINCIPLES

Sometimes measurements not on the blueprint are needed, and it may be necessary to subtract one fractional measurement from another to obtain the correct length of the materials.

Rule: To subtract fractions, reduce the fractions to the least common denominator and subtract the numerators only. Write the difference over the common denominator and reduce it to its lowest term. The whole numbers are subtracted in the usual manner.

Example: Subtract 1¾ from 3½

The fractions have unlike denominators so they must be changed to a least common denominator.

$$\frac{1}{2} \times \frac{2}{2} = \frac{2}{4}$$

$$\frac{3}{4} \times \frac{1}{1} = \frac{3}{4}$$

The problem now looks like this:

$$3\frac{2}{4}$$
$$-1\frac{3}{4}$$

Three-fourths cannot be subtracted from two-fourths, so we must borrow a whole number from the number 3, convert it to a fraction, and add it to ²⁄₄.

Subtract 1 (⁴⁄₄) from 3, and add it to the fraction ²⁄₄. 3²⁄₄ = 2 + ⁴⁄₄ + ²⁄₄ = 2⁶⁄₄

The result is 2⁶⁄₄. Now continue with subtraction.

$$2\frac{6}{4}$$
$$-1\frac{3}{4}$$
$$\overline{1\frac{3}{4}}$$

PRACTICAL PROBLEMS

1. Determine the missing dimension on this welded bracket. _____

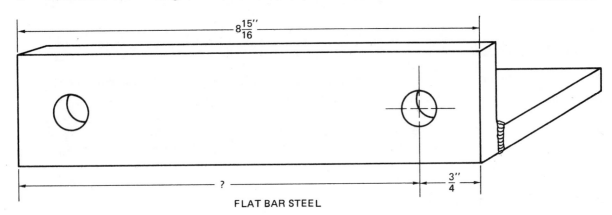

FLAT BAR STEEL

2. A 3¹⁄₁₆″ long piece is cut from the piece of steel angle iron illustrated. If there is ⅛″ of waste caused by the kerf of the oxy-acetylene cutting process, what is the length of the remaining piece of angle iron? _____

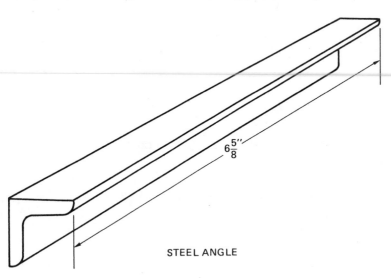

STEEL ANGLE

3. A 95⁄16″ long piece of bar stock is cut from this piece. What is the length
    of the remaining bar stock?

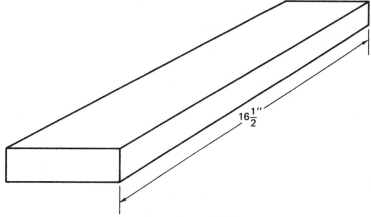

$16\frac{1}{2}''$

FLAT BAR STEEL

4. A 20¼″ diameter circle is flame cut from this square plate. Find the
    missing dimension. Subtract 1⁄16″ for waste.

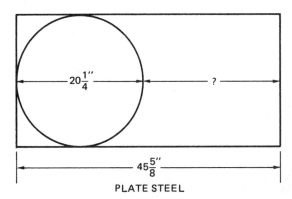

$20\frac{1}{4}''$?

$45\frac{5}{8}''$

PLATE STEEL

5. A square piece is flame cut from the center of a circle made from this
    steel plate. Find the missing dimension. Disregard waste caused by the
    width of the cut.

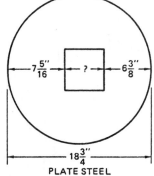

$7\frac{5}{16}''$? $6\frac{3}{8}''$

$18\frac{3}{4}''$

PLATE STEEL

6. A slot is cut from the center of a circle made from steel plate. Find the missing dimension. _____

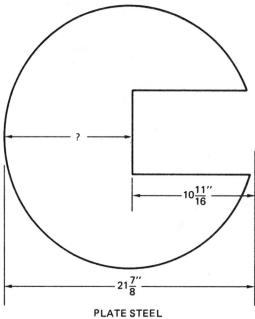

PLATE STEEL

7. A flame-cut wheel is to have the shape shown. Find the missing dimension. _____

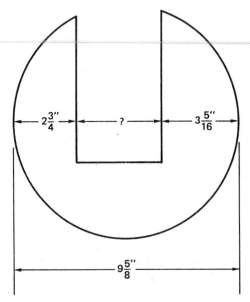

PLATE STEEL

Unit 8   MULTIPLICATION OF COMMON FRACTIONS

BASIC PRINCIPLES

The symbol for multiplication is "times" ($\times$). It is the short method of adding a number to itself a certain number of times.

Rule: To find the product of two or more fractions, multiply the denominators together and multiply the numerators together. Reduce the product to its lowest terms if necessary.

Example:

$$\frac{3}{4} \times \frac{7}{8} = \frac{21}{32}$$

This product cannot be reduced.

Cancellation Method:

$$\overset{3}{\underset{4}{\cancel{\frac{15}{16}}}} \times \overset{1}{\underset{1}{\cancel{\frac{4}{5}}}} = \frac{3}{4}$$

Cancel 5 into 15 (3 times) and 4 into 16 (4 times). The resulting fraction is ¾.

This product cannot be reduced.

When a combination of mixed numbers and fractions are to be multiplied, change the mixed number to an improper fraction, then multiply in the usual manner.

There is a very simple way to convert a mixed number to an improper fraction. Take our example (2½), for instance. First, multiply the whole number 2 by the denominator 2. The product is 4. Then, add that number (4) to the numerator 1. The total is 5. Since the original denominator is still 2, the fraction is 5⁄2.

Example: 2½ × ¼

$$= 2\frac{1}{2} \times \frac{1}{4} = 2 \times 2 + 1 = \frac{5}{2} \times \frac{1}{4}$$

$$= \frac{5}{8}$$

Let's try something a little more difficult.

Example: 2¾ × 1½

$$2\frac{3}{4} = 2 \times 4 + 3 = \frac{11}{4}$$

$$1\frac{1}{2} = 1 \times 2 + 1 = \frac{3}{2}$$

Then:

$$\frac{11}{4} \times \frac{3}{2} = \frac{33}{8} = 8\overline{)33} = 4\frac{1}{8}$$
$$\phantom{\frac{11}{4} \times \frac{3}{2} = \frac{33}{8} = 8\overline{)33}}\frac{32}{1}$$

PRACTICAL PROBLEMS

1.   This table needs four steel angle legs, each 20⅜″ long. How many inches of steel angle will be required to make the legs for 1 table?

a. _____

   For 20 tables?

b. _____

STEEL ANGLE

2. A welder has an order for 10 pieces of angle iron, each 9⅜″ long, and 21 pieces of I beam, each 16⁹⁄₁₆″ long. Answer a and b, allowing ⅛″ of waste for each saw cut.

_____

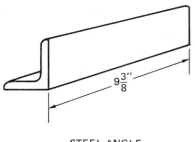

STEEL ANGLE   WIDE-FLANGE I BEAM

a. What is the total length of the steel angle required?

_____

b. What is the total length of the I beam required?

_____

3. Seven of these welded brackets are needed. What is the total length, in inches, of the bar stock needed for all of the brackets?

_____

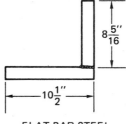

FLAT BAR STEEL

4. Sixty 3″ long pieces are cut from this bar of steel angle. There is ³⁄₁₆″ waste on each cut. What length of bar steel remains after the sixty pieces are cut?

_____

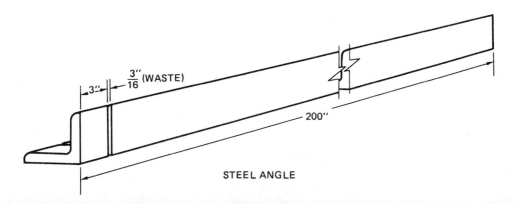

STEEL ANGLE

5.   Seven pieces of ½″ round stock, each 3″ long, are cut from a bar. How much material is required? Allow ⅛″ waste for each cut.   _____

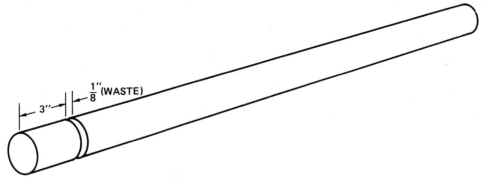

$\dfrac{1″}{8}$ (WASTE)

3″

ROUND BAR STOCK

6.   Thirteen pieces of steel angle, each 6⅞″ long, are welded to a piece of flat bar for use as concrete reinforcement. What is the total length of steel angle required?   _____

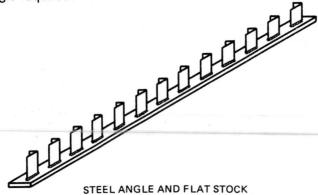

STEEL ANGLE AND FLAT STOCK

7.   To weld around this weldment, 16½ arc rods are needed. If 6¾ of the weldments are completed in an 8-hour shift, how many arc rods will be needed?   _____

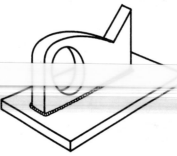

FLAT BAR STOCK

8.   This welded piece needs a total of 11⁷⁄₁₆″ of flat stock for reinforcement. If 13 of the weldments are required, how many inches of flat stock will be used for the completed order? Allow ⅛″ waste caused by the width of the cuts.

_____

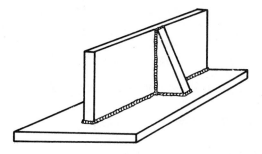

FLAT BAR STOCK

Unit 9   DIVISION OF COMMON FRACTIONS

BASIC PRINCIPLES

Rule: Invert the divisor, then multiply. (Invert means "Turn upside down." For example, $3/4''$ inverted is $4/3''$.)

Problem: How many $3\frac{1}{2}''$ shear pins may be cut from an 8' length of drill rod?

> Convert length of drill rod into inches.
> 8' x 12" (per foot) = 96".
> Divide 96" by the length of one shear pin:
> 96" divided by $3\frac{1}{2}''$ =
> Convert $3\frac{1}{2}''$ into a fractional equivalent, $3\frac{1}{2} = 7/2$.
> Invert the fraction to $2/7$.
> Multiply: 96" x $2/7$
>
> $$\frac{(96 \times 2)}{7} = \frac{192}{7} = 27.42 \text{ pins}$$
>
> The answer is 27 complete shear pins.

Example: $1/4''$ plate, 36" wide x 48" long.

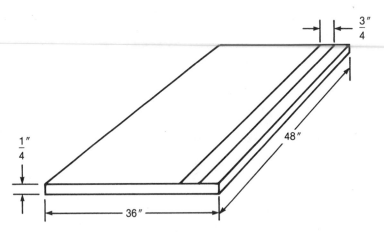

Cut $3/4''$ strips for engine mounts from this plate; each strip to be $3/4''$ wide x 48" long when finished, disregarding the steel that we will waste making the cut. If we cut all the strips possible, how many $3/4''$ strips will we have when the plate is completely used up?

Statement of the Problem: 36″ (plate) divided by ¾″ (strip);

$$\frac{36''}{1} \div \frac{3''}{4} = \frac{36''}{1} \times \frac{4''}{3} =$$

By cancellation, fractions are reduced to lowest terms. Fractions are then multiplied.

$$\frac{\overset{12}{\cancel{36}}}{1} \times \frac{4}{\cancel{3}_1} = 48 \text{ pieces.}$$

Proof: $\dfrac{\overset{12}{\cancel{48}}}{1} \times \dfrac{3}{\cancel{4}_1} = 36''$ (original size)

Division by mixed numbers:

12½ divided by 2½ =

25⁄2 divided by 5⁄2 =

25⁄2 x 2⁄5. By cancellation

50⁄10 = 5 Answer.

PROOF: 5 × 2 ½ = 12 ½

PRACTICAL PROBLEMS

1. A 28 ⅞″ piece of steel angle is in stock. How many 9³⁄16″ pieces may be cut from it? _____

2. It is necessary to cut as many keys as possible to fit this shaft. A piece of key stock, 15¾″ long, is in stock. How many keys may be sheared from it? Disregard waste caused by the width of the cuts. _____

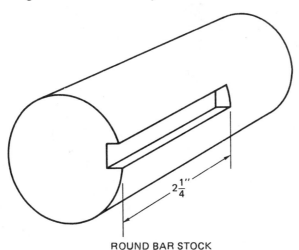

$2\frac{1}{4}''$

ROUND BAR STOCK

3.   This I beam is to be used for a sway-bar anchor. If the holes are equally
     spaced, what is the measurement between hole 1 and hole 2?   _____

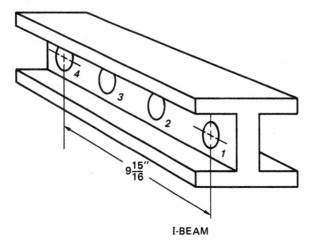

I-BEAM

4.   A bar of angle iron weighs 17¾ pounds. If 284 pounds of angle iron are
     in the stock pile, how many bars are in stock?   _____

5.   A piece of 16-gauge sheet metal 36″ wide is in stock. How many ¾″
     strips may be sheared from this sheet?   _____

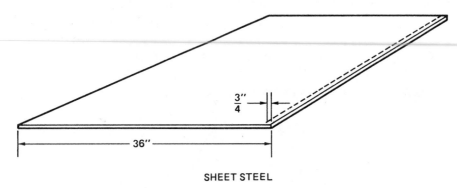

SHEET STEEL

6.   How many 2¼″ long pieces may be cut from a 14¾″ length of channel

7. Three bars of steel are shown. How many pieces, each 21½″ long, may be cut from the total length of the three bars after they are joined by welding?

_____

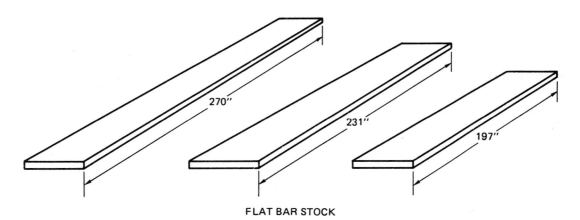

FLAT BAR STOCK

Unit 10 COMBINED OPERATIONS WITH COMMON FRACTIONS

BASIC PRINCIPLES

Review the principles of addition, subtraction, multiplication, and division of common fractions, and apply them to these problems.

Example: Find the dimensions on this welded cross.

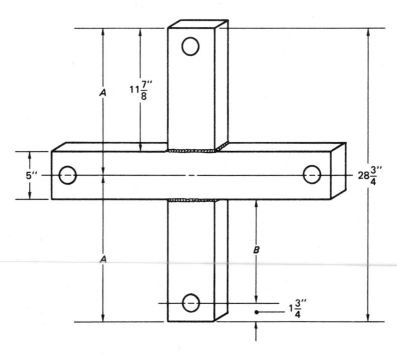

a. Dimension A equals 28¾" divided by 2 or

$$\frac{115''}{4} \text{ divided by 2} =$$

$$\frac{115}{8} = 14 \text{ ⅜" (Answer)}$$

b. Dimension B equals 11⅞" + 5" + 1¾" =

$$11⅞'' + 5'' + 16/8'' = 1713/8'' = 185/8'',$$

28¾" - 18⅝" = 28⁶/8" - 18⅝" = 10 ⅛"

PRACTICAL PROBLEMS

1. A piece of pipe has an inside diameter of 6″ and a wall thickness of ¼″.
    What is the outside diameter of the pipe?   _____

2. Twelve 24″ wide sections of steel are welded together. What is the total
    width of the steel after welding?   _____

3. Divide a 40″ sheet of metal into eight equal parts. What is the width of
    each part?   _____

4. A piece of mild steel flat stock is sheared and punched to the
    dimensions shown.

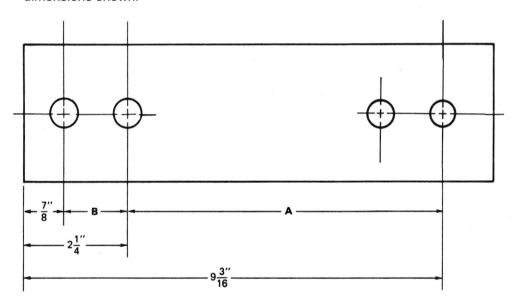

    a. Find dimension A.   a. _____

    b. Find dimension B.   b. _____

5.   Find dimension A on this steel angle.   _____

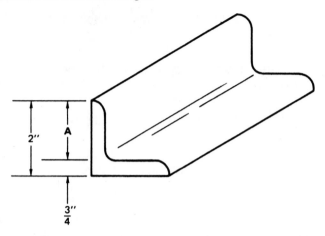

Decimal Fractions

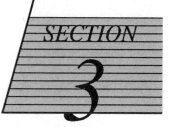

Unit 11   INTRODUCTION TO DECIMAL FRACTIONS

BASIC PRINCIPLES

Parts of inches may be stated in fractions ($\frac{1}{8}$", $\frac{1}{4}$", $\frac{3}{4}$") or in decimals 0.125", 0.25", 0.75".

If we divide a line into 10 parts, one part is one part divided by the total parts, or $\frac{1}{10}$ of the total.

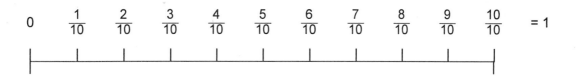

$$0 \qquad \frac{1}{10} \qquad \frac{2}{10} \qquad \frac{3}{10} \qquad \frac{4}{10} \qquad \frac{5}{10} \qquad \frac{6}{10} \qquad \frac{7}{10} \qquad \frac{8}{10} \qquad \frac{9}{10} \qquad \frac{10}{10} \qquad = 1$$

Two parts = $\frac{2}{10}$, three parts = $\frac{3}{10}$, and so on.

$\frac{10}{10}$ will cancel: $\frac{10}{10} = \frac{1}{1} = 1$ or the total line.

A *tenth* is written with a decimal point: $\frac{1}{10}$ = 0.1, $\frac{2}{10}$ 0.2. The first place to the right of a decimal point is called "tenths."

$$0.3 = \frac{3}{10} \qquad 0.4 = \frac{4}{10}$$

If one part ($\frac{1}{10}$) is 10 divided into 10 more parts, we have:

$$\frac{1}{10} \div \frac{10}{1} = \frac{1}{10} \times \frac{1}{10} = \frac{1}{100} \text{, written 0.01}$$

CHANGE FRACTIONS TO DECIMALS

Rule: Divide the numerator by the denominator. The answer will be on the *right* of the decimal point in tenths, hundredths, thousandths, and so on.

Problem: A piece of steel is $3\frac{1}{4}$" long. Convert the number to a decimal fraction.

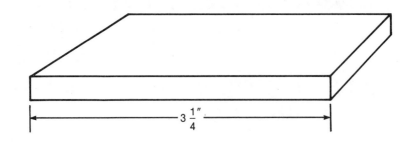

$$3\frac{1}{4}'' = 3'' + \frac{1}{4}''$$

Divide the numerator by the denominator:

$$\frac{1}{4}'' = 4\,\overline{)\,1}$$

Since 4 will not divide into 1, we must add a decimal point and zeros.

$$\begin{array}{r} .25 \\ 4\,\overline{)\,1.00} \\ \underline{.8} \\ .20 \end{array} \qquad \frac{1}{4}'' = 0.25 \text{ or } \frac{25}{100}$$

Answer: 3¼" = 325⁄100 or 3.25"

TO CHANGE DECIMALS TO FRACTIONS

The place value of the decimal becomes its denominator.

Example: 0.375 = (375⁄1000). Reduce to lowest terms if necessary.

$$\frac{375}{1000}$$

Reduced to lowest terms = 3⁄8.

To change a decimal to a fraction having a specific denominator, give the decimal fraction a denominator of 1, then multiply both the numerator and denominator by the desired denominator. Round off the numerator to a whole number if necessary.

Example: Change 0.875 to the nearest 32nd of an inch.

$$\frac{.875}{1} \times \frac{32}{32} = \frac{0.875}{1} \times \frac{32}{32} = \frac{28.000}{32} = \frac{28}{32}$$

ROUNDING OFF DECIMALS

If the last number in a decimal is greater than 5, the number may be *rounded off* by adding one to the preceding number.

Example: Rounding off to three decimal places: 14.3756 may be rounded off to 14.376; however, 14.3752 would be rounded off to 14.375, since the fourth number is less than 5.

Example: Rounding off to two decimal places: 14.376 may be rounded off to 14.38; however, 14.374 would be rounded off to 14.37 since the third number is less than 5.

If the fourth figure is *5*, leave the third figure unchanged if it is even; add 1 if it is odd.

$$14.3765 = 14.376 \text{ but } 14.3735 = 14.374$$

DECIMAL FRACTIONS

The second place to the right of the decimal point is called "hundredths."

$$0.3 = \frac{3}{10} \qquad 0.2 = \frac{2}{10}$$

$$0.03 = \frac{3}{100} \qquad 0.02 = \frac{2}{100}$$

The third place is called "thousandths."

$$\frac{1}{10} \times \frac{1}{10} \times \frac{1}{10} = \frac{1}{1000}$$

Each time we add another place to the *right* of the decimal, we multiply by $\frac{1}{10}$:

$$\frac{1}{10} = 0.1$$

$$\frac{1}{10} \times \frac{1}{10} = \frac{1}{100} = 0.01$$

$$\frac{1}{10} \times \frac{1}{10} \times \frac{1}{10} = \frac{1}{1000} = 0.001$$

$$\frac{1}{10} \times \frac{1}{10} \times \frac{1}{10} \times \frac{1}{10} = \frac{1}{10,000} = 0.0001$$

PRACTICAL PROBLEMS

1.   A length of channel is divided into 10 equal sections. Express each dimension as a decimal fraction.

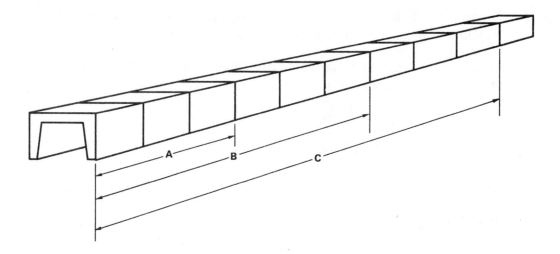

a.   Dimension A   a. _____

b.   Dimension B   b. _____

c.   Dimension C   c. _____

2.   Express each as a decimal number.

a.   three hundred seventy-five thousandths   a. _____

b.   one hundred twenty-five thousandths   b. _____

c.   six hundred twenty-five ten-thousandths   c. _____

d.   one thousand eight hundred seventy-five ten-thousandths   d. _____

e.   one tenth   e. _____

f.   seventy-five hundredths   f. _____

g.   one and zero tenths   g. _____

h.   ninety-six thousand eight hundred seventy-five hundred-thousandths   h. _____

3.   A welded tank holds 26.0478 gallons. Round this weight to the nearest thousandth gallon.   _____

4.   All of the following steel measurements may be given either in fractions or decimals. Find the decimal for each of these fractions:

Answer in
Decimal

a. $\frac{1}{16}$ inch   _____

b. $\frac{1}{8}$ inch   _____

c. $\frac{3}{16}$ inch   _____

d. $\frac{1}{4}$ inch   _____

e. $\frac{5}{16}$ inch   _____

f. $\frac{3}{8}$ inch   _____

g. $\frac{1}{2}$ inch   _____

h. $\frac{5}{8}$ inch   _____

i. $\frac{11}{16}$ inch   _____

j. $\frac{3}{4}$ inch   _____

Unit 12   ADDITION AND SUBTRACTION OF DECIMAL FRACTIONS

BASIC PRINCIPLES

Rule: To add or subtract decimals, place the numbers so that the decimal points are lined up one under the other, then add or subtract as you would whole numbers.

3.25	2.789	25.1	42.630	98.000
- 0.07	- 1.876	- 12.0	+ 18.275	+ 21.811
3.18	0.913	13.1	60.905	119.811

Note: When the numbers behind the decimal point are uneven, add zeros to reduce the chances of error.

When borrowing from whole numbers becomes necessary it may be accomplished in the following manner:

$$
\begin{array}{ccccc}
 & & & & 7\ 9\ 9 \\
98. & & 98.000 & & 9\cancel{8}.\cancel{0}\cancel{0}\cancel{0} \\
- 21.811 & = & - 21.811 & = & -21.811 \\
 & & & & 76.189
\end{array}
$$

PRACTICAL PROBLEMS

Note: Use this diagram for Problems 1–3.

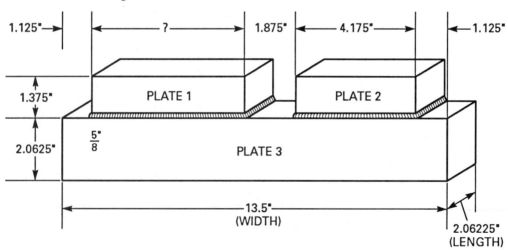

1.   Find the length of plate *1*.   _____

2.   An inventory of plate steel in a welding shop shows a balance of 16.375″
      of 2.0625″ square steel bar in stock. After the welder cuts plate *3* for the
      sliding block support, how much steel bar remains in stock?   _____

3.   The inventory shows a balance of 10.625″ of 1.375″ x 2.0625″ plate
      steel in stock. After the welder cuts plate *1* and plate *2* for the sliding
      block support, how much steel remains in stock?   _____

4.   a.   23.1   b.   65.4   c.   143.17
            - 12.2   - 32.65   - 22.16

      d. 198.3698   e.   5.25   f.   0.6347
          - 17.24   -0.8765   +0.5257

      g.   3.3146   h.   5.2   i.   2.52
          + 0.49   + 0.0198   + 1.36

      j.   2.64
            + .39

5.   What is the total weight of these three pieces of steel? Round the
      answer to two decimal places.   _____

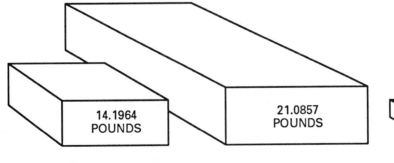
14.1964
POUNDS

21.0857
POUNDS

1.4689
POUNDS

6.   A welder saws a 3.5734″ long piece from this bar. Find, to the nearest thousandth inch, the length of the remaining bar. Allow a cutting loss of 0.125″.

_____

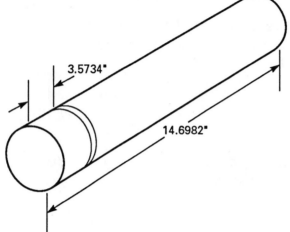

3.5734"

14.6982"

Unit 13   MULTIPLICATION OF DECIMAL FRACTIONS

BASIC PRINCIPLES

The multiplication of decimals is used to find the total length, width, or height of materials for specific jobs. If the length is in decimals and we have a certain number of materials, we can arrive at their overall length, width, or height by simple multiplication of the decimals.

Example: A piece of angle iron 15.685 inches long is needed 13 times on a frame repair job. How many inches of angle iron must be available for the job?

$$
\begin{array}{r}
15.685 \\
\times \quad 13 \\
\hline
47055 \\
15685 \quad \\
\hline
203.905
\end{array}
$$

    15.685   (3 decimal places)

  x   13   (0 decimal places)

    47055

  15685

  203.905   (3 decimal places in answer)

     | | |

     321   (Start counting places from right side.)

Remember: The number of decimal places is counted in the multiplicand and multiplier, then added. This determines the number of decimal places that are in the product (answer).

    6.875   (3 decimal places)

  x 0.501   (3 decimal places)

    6875

  343750

  3.444375   (6 decimal places)

  | | | | | |

  654321

But:

    .15   (2 decimal places)

  x .02   (2 decimal places)

  .0030   (4 decimal places)

The answer has only two numbers in it, so we must add zeros to the answer to have a total of four decimal places (.0030). The zeros must be added in front of the product.

PRACTICAL PROBLEMS

1. Nineteen pipe flanges are flame cut from 16.25″ wide plate. How much
    plate is required for the 19 flanges? Assume that there is no waste due
    to cutting.

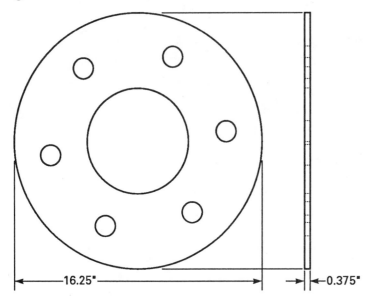

|←—————16.25″—————→| →| |←0.375″

2. These 19 flanges are stacked as shown. Each flange is 0.375″ high.
    What is the total height of the pile?

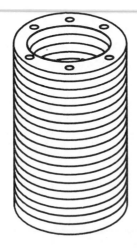

3. A welder uses 3.1875 cubic feet of acetylene gas to cut one flange. How
    much acetylene gas is used to cut the 19 flanges?

4. A welder uses 6.968 cubic feet of oxygen gas to cut one flange. How much oxygen gas is used to cut the 19 flanges? _____

Note: Use this diagram for Problems 5–7.

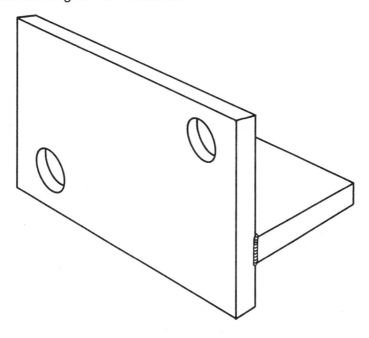

5. Each of these welded brackets weighs 2.827 pounds. A welder makes 13 of the brackets. What is the total weight of the 13 brackets? _____

6. The steel plate used to make the brackets cost 0.215 dollar per pound. Each bracket weighs 2.827 pounds. What is the total cost of the order of 13 brackets? Round the answer to the nearest whole cent. _____

7. The welder cuts two holes in each bracket. Each bolt-hole cut wastes 0.1875 pound of material. Find, in pounds, the amount of waste for the order of 13 brackets. _____

8. A welder cuts 14 squares from a piece of plate. Each side is 4.606″. What is the total length of 4.606″ wide stock needed? Round the answer to two decimal places. Disregard waste caused by the width of the cuts. _____

9. A welder cuts 6 pieces of steel angle, each 3.0698″ long. The total length is rounded to the nearest thousandth inch. What is the total length? Disregard waste caused by the width of the cuts. _____

 Unit 14   DIVISION OF DECIMAL FRACTIONS

BASIC PRINCIPLES

To divide decimal fractions, set up the problem as for regular division, but remember to carry decimal points in the answer.

Example: Divide 20.5 by 4.

```
        5.125
   4 ) 20.50
       20
        5
        4
       10
        8
       20
       20
       00
```

20.5 divided by 4 = 5.125. The decimal point always appears directly above the decimal point in the problem.

Example: A piece of frame material is 18½ feet long. If we cut it into 9 equal pieces, how long will each piece be?

Using decimals: 18½ feet = 18.5 feet.

```
        2.05
   9 )18.50
      18
       50
       45
        5      (remainder)
```

Ignoring the remainder, each piece will be 2.05 feet long.

Example: Divide 15.3 by 0.03.

If the divisor is a decimal fraction, move the decimal point to the right of the number until it is a whole number: 0.03 = 3 (two places).

The decimal point in the dividend must also be moved the same number of places to the right: 15.3 = 1530 (two places).

Zeros must be added if there are not enough digits in the dividend.

$$0.03 \,\overline{)15.3} = 3. \,\overline{)1530.} = 3 \,\overline{)1530}$$

$$
\begin{array}{r}
510 \\
3\,\overline{)1530} \\
\underline{15} \\
3 \\
\underline{3} \\
0
\end{array}
$$

Proof:

$$
\begin{array}{r}
510 \\
\times\,.03 \\
\hline
15.30
\end{array}
$$

PRACTICAL PROBLEMS

Note: Use this diagram for Problems 1–4

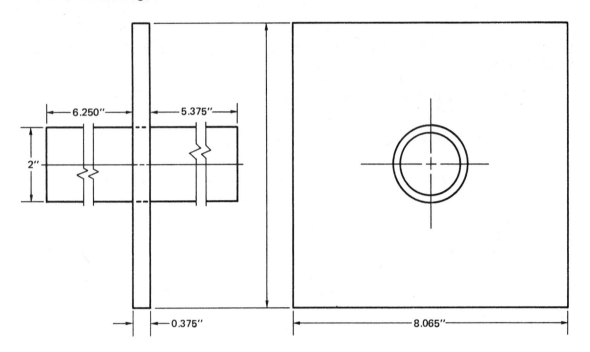

1.   The pipe on the connection socket weighs 1.375 pounds per foot (12 inches). How many connection sockets can be made from an inventory of 31.625 pounds of pipe? Disregard waste caused by the width of the cuts made by the saw blade.   _____

2.   How many feet of pipe are in the inventory?   _____

3.   How many baseplates for the connection socket can be cut from a socket plate 185.495″ x 185.495″? Allow 3⁄16″ cutting waste.   _____

4.   The 23 waste blanks weigh 9.6899 pounds. How much does each blank weigh?   _____

5. A welder saw cuts this length of steel angle into seven equal pieces. What is the length of each piece? Disregard waste caused by the width of the cuts. Round the answer to four decimal places.

_____

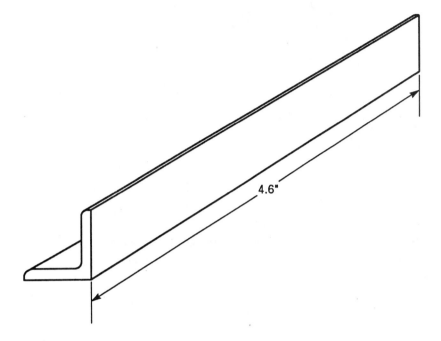

4.6"

6. A welder shears this plate into pieces that are 9″ wide. How many whole pieces are sheared? Disregard waste caused by the width of the cuts.

_____

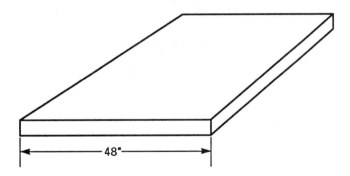

48"

7. A welder shears key stock into pieces 2.75″ long. How many whole pieces are sheared from a length of key stock 74.15″ long? Disregard waste caused by the width of the cuts.

_____

8. The welder saws this square stock into four equal pieces and rounds the length of the cut pieces to the nearest hundredth inch. What is the length of each piece? Allow 1⁄16″ waste for each cut.

_____

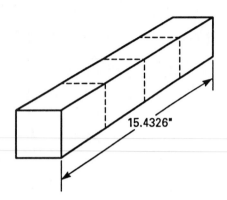

15.4326″

9. The welder flame cuts this plate into seven equal pieces, each 60 inches long. Find, to the nearest hundredth inch, the width of each piece. Allow 3⁄16″ waste for each cut.

_____

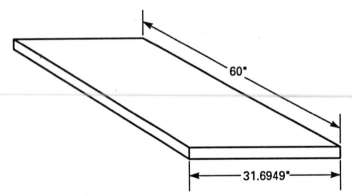

60″

31.6949″

 # Unit 15   DECIMAL AND COMMON FRACTION EQUIVALENTS

BASIC PRINCIPLES

Review the principles of decimal and common fraction equivalents and apply them to these problems.

Review denominate numbers in Section 1 of the Appendix.

Example: Convert the fraction 1/4 to a decimal as follows: $1/4 = 4\overline{)1}$

$$
4\overline{)1.00} = \quad
\begin{array}{r}
.25 \\
4\overline{)1.00} \\
\underline{8} \\
20 \\
\underline{20} \\
0
\end{array}
$$

or:

$$
1/4 \;=\; \frac{25}{100} \;=\; 0.25
$$

To change a decimal to a given fraction, multiply the decimal by the given fraction and place the answer over the given fraction.

Example: Change the decimal 0.375 to 16ths.

$$
\begin{array}{r}
0.375 \\
\times\ 16 \\
\hline
2250 \\
375 \\
\hline
6.000
\end{array}
$$

Then:

$$
\frac{6}{16} = 0.375
$$

PRACTICAL PROBLEMS

1. Express the fractional inches as a decimal number.

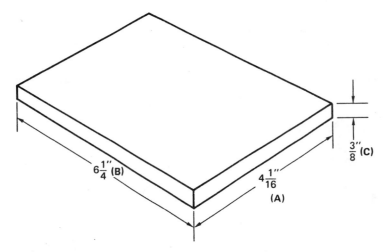

a. Dimension A   a. _____

b. Dimension B   b. _____

c. Dimension C   c. _____

2. Express each dimension in feet and inches. Express the inches as a decimal number.

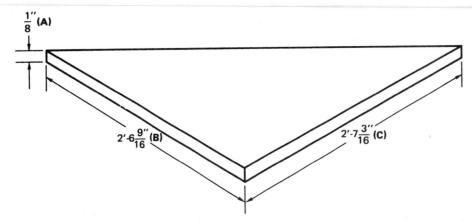

a. Dimension A   a. _____

b. Dimension B   b. _____

c. Dimension C   c. _____

3. Express each decimal dimension as a fractional number.

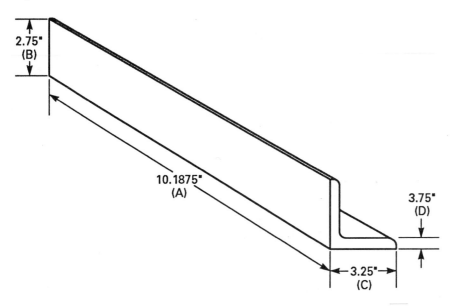

    a. Dimension A   a. _____

    b. Dimension B   b. _____

    c. Dimension C   c. _____

    d. Dimension D   d. _____

4. Express each dimension as a fractional number. Reduce the answers to lowest terms.

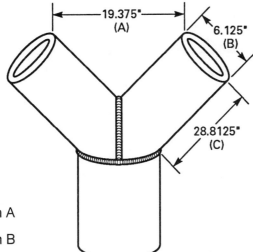

    a. Dimension A   a. _____

    b. Dimension B   b. _____

    c. Dimension C   c. _____

5.   Express each dimension as a fractional number.

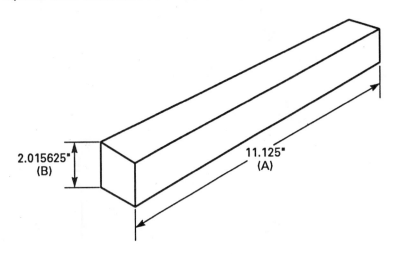

   a.   Dimension A   a. _____

   b.   Dimension B   b. _____

6.   Express each decimal as indicated.

   a.   0.375 inch to the nearest 32nd inch   a. _____

   b.   0.0625 inch to the nearest 16th inch   b. _____

   c.   0.9375 inch to the nearest 16th inch   c. _____

   d.   0.625 inch to the nearest 8th inch   d. _____

   e.   0.750 inch to the nearest 4th inch   e. _____

Note: Use this table for Problems 7–12.

DECIMAL EQUIVALENT TABLE

Fraction	Decimal	Fraction	Decimal	Fraction	Decimal	Fraction	Decimal
1/64	0.015625	17/64	0.265625	33/64	0.515625	49/64	0.765625
1/32	0.03125	9/32	0.28125	17/32	0.53125	25/32	0.78125
3/64	0.046875	19/64	0.296875	35/64	0.546875	51/64	0.796875
1/16	0.0625	5/16	0.3125	9/16	0.5625	13/16	0.8125
5/64	0.078125	21/64	0.328125	37/64	0.578125	53/64	0.828125
3/32	0.09375	11/32	0.34375	19/32	0.59375	27/32	0.84375
7/64	0.109375	23/64	0.359375	39/64	0.609375	55/64	0.859375
1/8	0.125	3/8	0.375	5/8	0.625	7/8	0.875
9/64	0.140625	25/64	0.390625	41/64	0.640625	57/64	0.890625
5/32	0.15625	13/32	0.40625	21/32	0.65625	29/32	0.90625
11/64	0.171875	27/64	0.421875	43/64	0.671875	59/64	0.921875
3/16	0.1875	7/16	0.4375	11/16	0.6875	15/16	0.9375
13/64	0.203125	29/64	0.453125	45/64	0.703125	61/64	0.953125
7/32	0.21875	15/32	0.46875	23/32	0.71875	31/32	0.96875
15/64	0.234375	31/64	0.484375	47/64	0.734375	63/64	0.984375
1/4	0.250	1/2	0.500	3/4	0.750	1	1.000

7.   A welder cuts these four pieces of metal. Express each dimension as a fractional number.

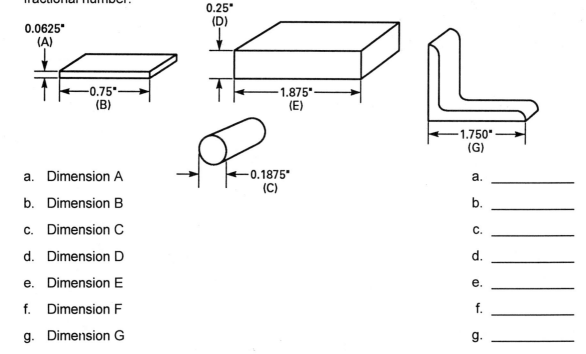

a.   Dimension A

b.   Dimension B

c.   Dimension C

d.   Dimension D

e.   Dimension E

f.   Dimension F

g.   Dimension G

a. _____

b. _____

c. _____

d. _____

e. _____

f. _____

g. _____

8. A piece of steel channel and a piece of I beam are needed. Express each dimension as a decimal number.

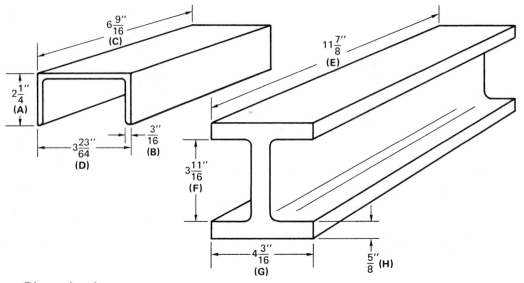

a. Dimension A   a. _____

b. Dimension B   b. _____

c. Dimension C   c. _____

d. Dimension D   d. _____

e. Dimension E   e. _____

f. Dimension F   f. _____

g. Dimension G   g. _____

h. Dimension H   h. _____

9. Express each decimal as a fraction.

a. 0.171875   a. _____

b. 0.3125   b. _____

c. 0.875   c. _____

d. 0.5625   d. _____

e. 0.9375   e. _____

f. 0.515625   f. _____

10.   Express each fraction as a decimal.

a.   ¾   a.   _____

b.   31⁄32   b.   _____

c.   23⁄32   c.   _____

d.   5⁄8   d.   _____

e.   ½   e.   _____

f.   3⁄8   f.   _____

g.   3⁄16   g.   _____

ROUNDING OFF DECIMALS—A REVIEW METHOD

When working with decimals, there is sometimes a remainder. When you have a remainder, it is either added on the number or dropped. This is called "rounding off."

Generally, if a remainder is less than *5* it is dropped; if the remainder is *5* or over, *1* is added on.

Example:

We have a number:

5.254

The third decimal is "4" and it is less than 5 so we drop it.

Now our number is

5.25

This (5.25) is the number we use. But suppose our number is

5.256

The third decimal is "6" and it is more than 5 so we add one to the number ahead of it.

Now our number is

5.26

This (5.26) is the number we use.

Unit 16   TOLERANCES

BASIC PRINCIPLES

Tolerance is the amount over or under a given measurement that is allowed for a fit-up.

The given dimension *plus* the tolerance gives the longest possible length (or thickness). The dimension *minus* the tolerance gives the shortest possible length (or thickness).

Example:

$$A = \begin{array}{r} 18.1250 \\ + \ 0.0625 \\ \hline 18.1875 \end{array}$$

$$B = \begin{array}{r} 18.1250 \\ - \ 0.0625 \\ \hline 18.0625 \end{array}$$

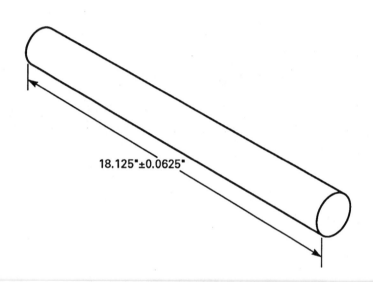

18.125"±0.0625"

PRACTICAL PROBLEMS

Using the given tolerances, find A, the longest possible length, and B, the shortest possible length.

1.

A = _____

B = _____

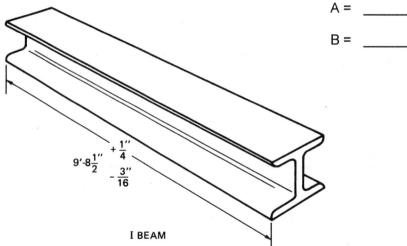

$9'\text{-}8\frac{1}{2}'' \quad \begin{array}{l} +\frac{1''}{4} \\ -\frac{3''}{16} \end{array}$

I BEAM

2.

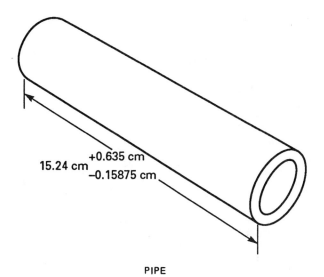

PIPE

A = _____

B = _____

3.

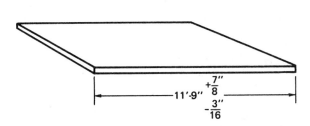

RECTANGULAR STEEL PLATE

A = _____

B = _____

4.

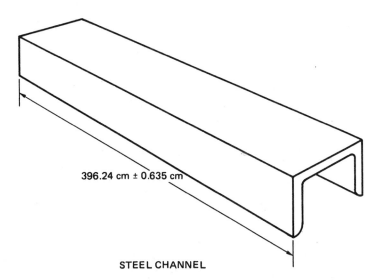

STEEL CHANNEL

A = _____

B = _____

5.

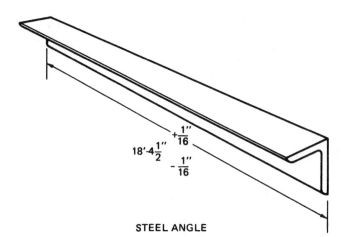

18'-4$\frac{1}{2}''$ $+\frac{1''}{16}$ $-\frac{1''}{16}$

STEEL ANGLE

A = _____

B = _____

6.

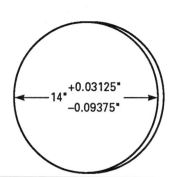

14" +0.03125" −0.09375"

A = _____

B = _____

7.

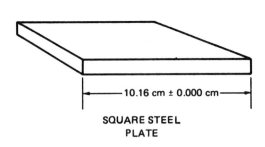

10.16 cm ± 0.000 cm

SQUARE STEEL
PLATE

A = _____

B = _____

Unit 17   COMBINED OPERATIONS WITH DECIMAL FRACTIONS

BASIC PRINCIPLES

Review the principles of addition, subtraction, multiplication, and division of decimal fractions, and apply them to these problems.

Example: If A = 3.25″, C = 3.25″, and the total width of A, B, and C = 8.5625″, find the diameter of B.

$$
\begin{array}{ll}
3.25 & \text{(A)} \\
+\ 3.25 & \text{(C)} \\
\hline
6.50 & \text{(Total of A + C)}
\end{array}
$$

Then:

$$
\begin{array}{ll}
8.5625 & \text{(Total width)} \\
-\ 6.50 & \\
\hline
2.0625 & \text{(Total width of B)}
\end{array}
$$

Proof:

$$
\begin{array}{l}
3.2500 \\
3.2500 \\
+\ 2.0625 \\
\hline
8.5625
\end{array}
$$

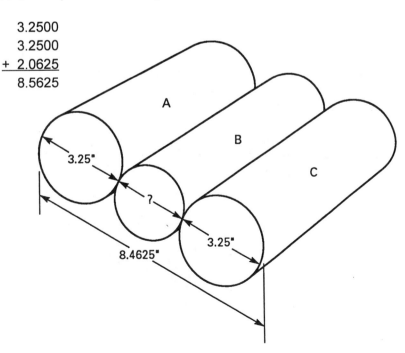

PRACTICAL PROBLEMS

1. Find the length of slot 2. _____

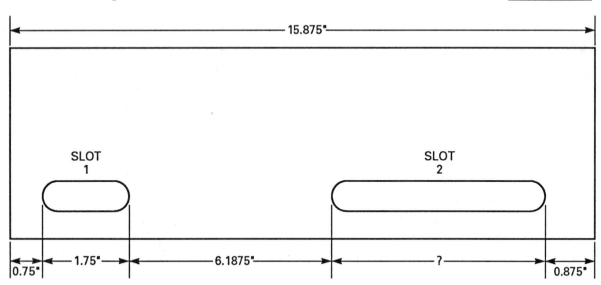

2. Find the height of a stack of 13 of these steel shims. _____

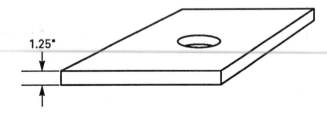

3. Cross-bar members are cut from flat stock. What length of 5″ flat stock is used to make 31 of these members? Disregard waste caused by the width of the cuts. _____

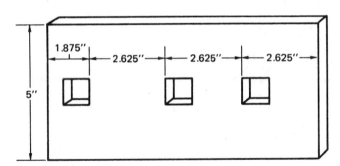

4. The round stock shown here is sawed into 3.5″ pieces. How many 3.5″ pieces can be made? Allow 0.0625″ waste for each cut. _____

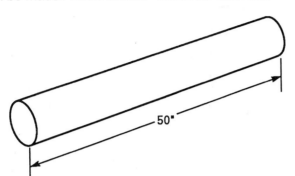

—50ʺ—

5. A welded truck-bed side support is shown. How many complete supports can be cut from a length of 2″ x 127.875″ long bar stock? Disregard waste caused by the width of the cuts. _____

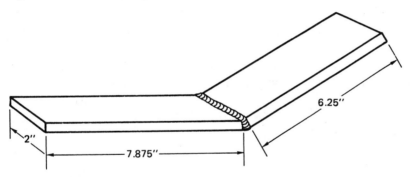

6.25″

2″

—7.875″—

6. Convert the length of this steel channel to three decimal places. _____

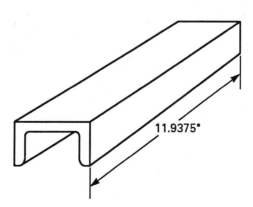

11.9375ʺ

7. Convert the length of the pipe on this welded support to two decimal places.

_____

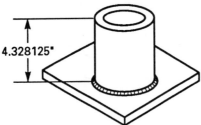

4.328125"

8. Use the decimal equivalent table to express each fraction as a decimal.

a. 33⁄64

b. 7⁄64

c. 5⁄16

d. 7⁄8

a. _____

b. _____

c. _____

d. _____

9. What is the thickness of one wall of a pipe that has the inside diameter of 15.72 centimeters and the outside diameter of 16.50 centimeters?

_____

10. A welded bracket has the dimensions shown. Find dimension A.

_____

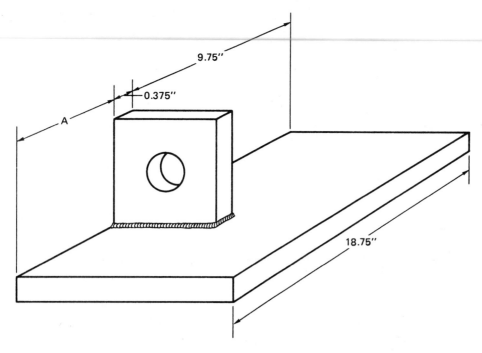

9.75"

0.375"

A

18.75"

Averages, Percents, and Percentages

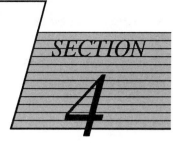

SECTION

4

Unit 18   AVERAGES

BASIC PRINCIPLES

To find the average of two or more quantities, they are first added together. Then, this sum is divided by the number of quantities.

Example:

Find the *average* of ¼", ³⁄₁₆", ½", and ⁵⁄₁₆".

Place the numbers in a column for addition:

$$\frac{1}{4}$$

$$\frac{3}{16}$$

$$\frac{1}{2}$$

$$+\ \frac{5}{16}$$

Since the fractions cannot be added in this form, we must find the common denominator. As in any case of addition of shop fractions, we use the largest denominator, which in this case is *16*.

Therefore, we will reduce all the fractions to sixteenths:

$$\frac{1}{4} = \frac{4}{16}$$

$$\frac{3}{16} = \frac{3}{16}$$

$$\frac{1}{2} = \frac{8}{16}$$

$$+\ \frac{5}{16} = +\ \frac{5}{16}$$

$$\frac{20}{16} \qquad \frac{20''}{16} = \text{Total}$$

69

There are four numbers. To find the average of the four numbers, we divide the total of all the numbers by the number of quantities (4).

$$\frac{20}{16} \div \frac{4}{1} =$$

$$\frac{20}{16} \times \frac{1}{4} =$$

$$\frac{\overset{5}{\cancel{20}}}{16} \times \frac{1}{\underset{1}{\cancel{4}}} =$$

$$\frac{5 \times 1}{16 \times 1} = \frac{5}{16}$$

Answer: The average of ¼", ³⁄16", ½", and ⁵⁄16" = ⁵⁄16".

PRACTICAL PROBLEMS

1. Six pieces of steel angle are cut to make a weldment. What is the average length of the pieces? Round the answer to three decimal places.

_____

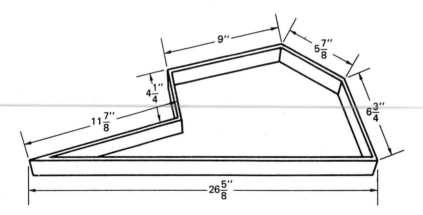

2. Five pieces of 1″ square bar stock are cut as shown. What is the average length, in inches, of the pieces? _____

3. Six welding jobs are completed using 33 electrodes, 19 electrodes, 48 electrodes, 14 electrodes, 31 electrodes, and 95 electrodes. What is the average number of electrodes used for each job? _____

4. On 5 jobs a welder earns $14.56, $23.98, $15.00, $32.40, and $17.58. Find, to the nearer whole cent, the average earned per job. _____

5. A piece of steel plate is measured for thickness in different places. These measurements are found: 1 ¼″, 1 ³⁄₁₆″, 1 ¼″, 1 ½″. What is the average thickness of the plate? Round the answer to the nearest thousandth. _____

6. Four pieces of ½″ plate weigh 10 pounds, 26 pounds, 9¾ pounds, and 29½ pounds. Find the average weight of the plates to the nearest thousandth pound. _____

7. Six plates are stacked and weighed. The total weight is 210 pounds. What is the average weight of each piece? _____

8. A welded steel tank holds 325 gallons. Another tank holds twice as much. What is the average amount held by the tanks? _____

 # Unit 19   PERCENTS AND PERCENTAGES

BASIC PRINCIPLES

Definition: A percent of a number is a method of expressing some part of whole numbers with a base of 100; for example, 100% of a number is all of it.

Example: Express ½ and ¼ as percents.

$$\frac{1}{2} = .50 = 50\%$$

$$\frac{1}{4} = .25 = 25\%$$

Note that the fraction is changed to a decimal by dividing the numerator by the denominator first. Then the decimal point is moved two places right, dropped, and the percent sign is added.

Proof: $\frac{50}{100} = \frac{1}{2}$

$\frac{25}{100} = \frac{1}{4}$

0	$\frac{1}{4}$	$\frac{1}{2}$	$\frac{3}{4}$	1	Fraction
0	25%	50%	75%	100%	Percent

A whole number appears on the *left* side of a decimal point; parts of a whole number appear on the *right* side of the decimal point.

1	6	9	12	(Whole numbers)
.25	.20	.18	.40	(Parts)
1.25	6.20	9.18	12.40	(Whole numbers and parts)

One dollar is written:

Decimal point

$1.00

Whole number   Decimal or part

One dollar and 25 cents is written: $1.25

Part of 1 dollar

72

All percentage problems consist of three elements: (a) the *base,* (b) the *rate,* and (c) the *percentage.*

Given the problem, 5% of $100.00 is $5.00, the *base* is $100.00, the *rate* is 5, and the *percentage* is $5.00.

Example: If the *base* and *rate* are given, the *percentage* is found by multiplying the base by the rate.

Problem: Find 5% of $100.00:

$100
x 0.05   (2 decimal places)
$5.00   (2 decimal places)

Note: The percentage symbol (%) does the work of *two* decimal places. 5% is written 0.05; 18% is written 0.18, and so on. 125% is written as 1.25 (two decimal places), but 3¼% is written as 0.0325 since the *25* is a quarter of the whole number 3, which is written 0.03.

Example: If the *base* and the *percentage* are given, divide the base into the percentage to find the *rate.*

Problem: What % of $100.00 is $5.00?

$$\frac{\$5}{\$100} = 100 \overline{)\begin{array}{r} 0.05 \quad = 5\% \\ 5.00 \\ \underline{5\ 00} \\ 0 \end{array}}$$

Example: If the *percentage* and *rate* are given, divide the rate into the percentage to find the *base.*

Problem: 5 is 5% of what number?

$$\frac{5}{0.05} = 0.05 \overline{)\begin{array}{r} 100 \\ 5.00 \\ \underline{5.00} \\ 000 \end{array}}$$

RULES OF CONVERSION

Example: To change a fraction to a percent, multiply by 100.

Problem:

$$\frac{3}{8} \times \frac{100}{1} = 8\overline{)300} \quad \begin{array}{r} 37 \\ \hline 300 \\ 24 \\ \hline 60 \\ 56 \\ \hline 4 \end{array}$$

$$\frac{4}{8} = \frac{1}{2}$$

Note that the remainder is 4. That remainder is placed over the divisor (8). This creates the fraction 4/8. Reducing the fraction to its lowest terms, we get 1/2. Combining the whole number (37) with the fraction (1/2) and adding the percent sign gives the answer 37½%.

Example: To change a % to a fraction, drop the % sign and multiply by 100.

Problem: 37 ½% x 100

$$37\frac{1}{2} \times \frac{1}{100} = \frac{75}{2} \times \frac{1}{100} = \frac{75}{200} = \frac{3}{8}$$

Answer: 37 ½% = ⅜

Example: To change decimals to percents, move the decimal point two places to the right.

Problem: Change 0.20 to a percent.

Answer: 0.20 = 20%

Example: To change percents to decimals, move the decimal point two places to the left.

Problem: Change 20% to a decimal.

Answer: 20% = .20

PRACTICAL PROBLEMS

1. Express each percent as a decimal.

    a. 1%

    b. 5%

    c. 8%

    d. 60%

    e. 23¼%

    f. 125%

    g. 220%

    a. _____

    b. _____

    c. _____

    d. _____

    e. _____

    f. _____

    g. _____

2. A welder works 40 hours and earns $16.00 per hour. The deductions are: income tax, 18%; Social Security, 7%; union dues, 5%; hospitalization insurance, 2½%. Find each amount to the nearest whole cent.

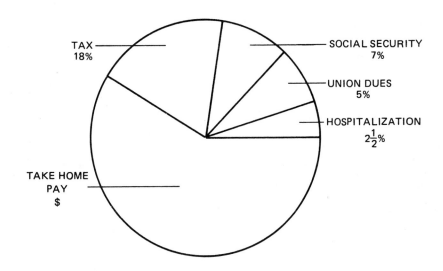

    a. hospitalization insurance

    b. income tax

    c. Social Security

    d. union dues

    e. net, or take home, pay

    a. _____

    b. _____

    c. _____

    d. _____

    e. _____

3. The area of a piece of steel is 1446.45 square inches. How many square inches are contained in 25% of the steel?

    _____

4. A welder completes 87% of 220 welds. How many completed welds are made?

    _____

5. A total of 116,000 welds are made in a welding shop. This graph of the quality of the welds is made.

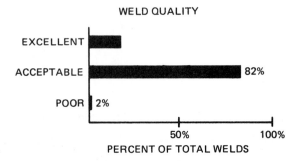

WELD QUALITY

PERCENT OF TOTAL WELDS

a. How many welds are of poor quality?   a. _____

b. How many welds are acceptable?   b. _____

c. How many welds are of excellent quality?   c. _____

6. In a mill, 10,206 steel plates are sheared. By inspection, 20% of the plates are rejected. Of the amount that are rejected, 8% are scrapped.

a. How many complete steel plates are rejected?   a. _____

b. How many of the rejected plates are scrapped? b. _____

Direct Measure

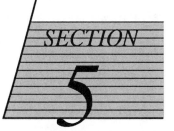

Unit 20   DIRECT MEASURE INSTRUMENTS

BASIC PRINCIPLES

Review the principles of direct measure instruments and apply them to these problems.

Note: Use this information for Problems 1–5.

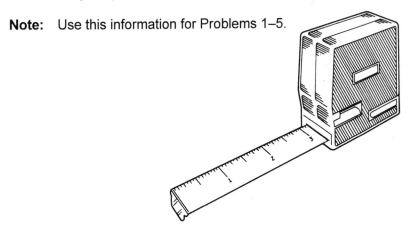

Generally speaking, most tapes used by welders are of the 6-foot variety, although at times it is more convenient to use tapes that are longer. The most convenient are graduated in $\frac{1}{32}$″, $\frac{1}{16}$″, $\frac{1}{8}$″, $\frac{1}{4}$″, $\frac{1}{2}$″. It is most important that the welder learn to read a tape, and is necessary for accuracy in measuring work. Some tapes even give $\frac{1}{64}$″ spaces, but rarely can a cut be made so close to such a measurement. Cuts that need to be close to fine measurements will generally be made with a saw or shear.

PRACTICAL PROBLEMS

1.   Read and record the distance from the start of this steel tape measure to the letter A.   A = _____

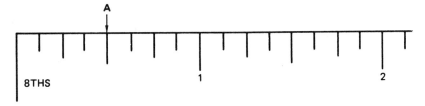

2.   Read the distances from the start of this steel tape measure to the
     letters. Record the answers in the proper blanks.   B = _____

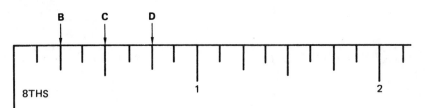

                                                                           C = _____

                                                                           D = _____

3.   Read the distances from the start of this steel tape measure to the
     letters. Record the answers in the proper blanks.   E = _____

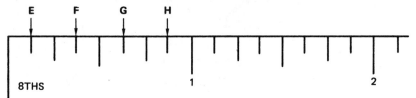

                                                                           F = _____

                                                                           G = _____

                                                                           H = _____

4.   Read the distances from the start of this steel tape measure to the
     letters. Record the answers in the proper blanks.   I = _____

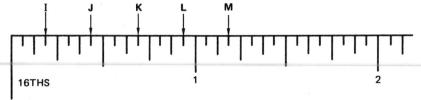

                                                                           J = _____

                                                                           K = _____

                                                                           L = _____

                                                                           M = _____

5.   Using a ruler, draw lines of these lengths and label your drawings with
     the correct measurements.

     a. $2\frac{1}{8}$ in

     b. $3\frac{3}{8}$ in

     c. $1\frac{1}{4}$ in

     d. $1\frac{3}{4}$ in

     e. $1\frac{1}{2}$ in

     f. $1\frac{15}{16}$ in

     g. $1\frac{15}{16}$ in

     h. $3\frac{1}{2}$ in

 # Unit 21   METRIC LENGTH MEASURE

BASIC PRINCIPLES

Review denominate numbers in Section I of the Appendix.

Study this table of metric length measure.

METRIC LENGTH MEASURE

10 millimeters (mm)	=	1 centimeter (cm)
10 centimeters (cm)	=	1 decimeter (dm)
10 decimeters (dm)	=	1 meter (m)
10 meters (m)	=	1 dekameter (dam)
10 dekameters (dam)	=	1 hectometer (hm)
10 hectometers (hm)	=	1 kilometer (km)

Study these principles of equivalent measures of denominate numbers.

To express a metric length unit as a smaller metric length unit, multiply by 10, 100, 1,000, 10,000, and so forth.

To express a metric length unit as a larger metric length unit, multiply by 0.1, 0.01, 0.001, 0.0001, and so forth.

Example: This bar is 10 centimeters long. What is the length in millimeters? See the metric length measure table above.

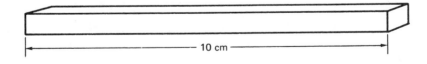

10 cm

     10 mm = 1 cm
     10 cm = 100 mm

Answer: 100 mm

PRACTICAL PROBLEMS

1. This piece of steel channel has a length of 22 millimeters. Express this
    measurement in centimeters.   _____

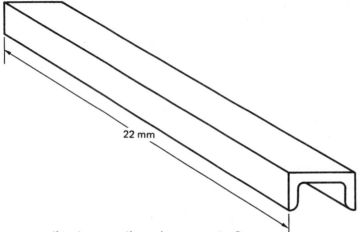

22 mm

2. How many centimeters are there in one meter?   _____

3. A pipe with end plates is shown.

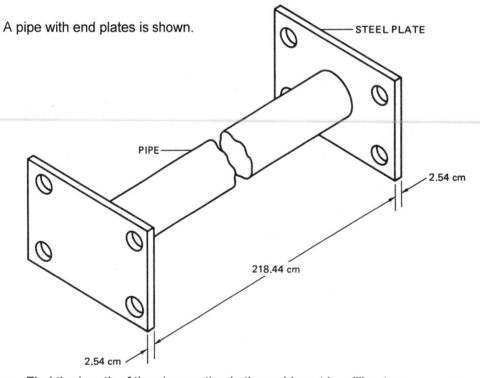

STEEL PLATE

PIPE

2.54 cm

218.44 cm

2.54 cm

a. Find the length of the pipe section in the weldment in millimeters.   a. _____

b. Find the thickness of one end plate in millimeters.   b. _____

c. Find the overall length in meters.   c. _____

Note: Use this diagram for Problems 4–7.

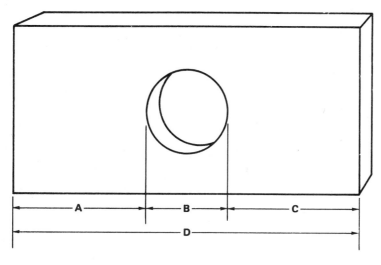

FLAT BAR

4.   A = 36 centimeters, B = 36 centimeters, and D = 108 centimeters. Find dimension C.

_____

5.   B = 48 centimeters, C = 56 centimeters, and D = 122 centimeters. Find dimension A.

_____

6.   C = 56 millimeters, D = 160 millimeters, and A = 56 millimeters. Find dimension B.

_____

7.   B = 68 centimeters, C = 52 centimeters, and A = 98 millimeters. Find dimension D.

_____

8.   This piece of bar stock is cut into pieces, each 10 centimeters long. How many pieces are cut?

_____

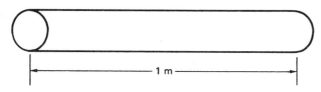

1 m

The metric steel rule is used to measure various lengths of steel in metric units. Metric measurements are expressed in decimal parts of a whole number. For example, one-half millimeter is written as 0.5 mm.

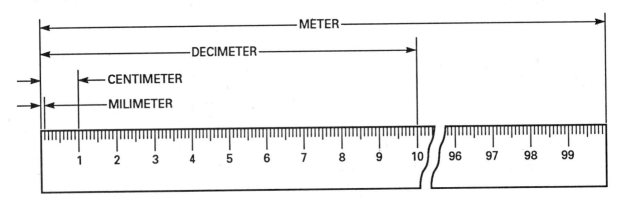

Note: Use this diagram for Problems 9 and 10.

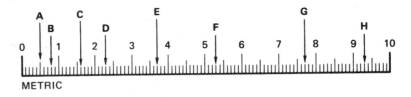

METRIC

9. Read the distances, in millimeters, from the start of the rule to the letters on the rule. Record the answers in the proper blanks.

A = _____

B = _____

C = _____

D = _____

E = _____

F = _____

G = _____

H = _____

10. Read the distances, in centimeters, from the start of the rule to the letters on the rule. Record the answers in the proper blanks.

A = _____

B = _____

C = _____

D = _____

E = _____

F = _____

G = _____

H = _____

11. A shaft support is shown.

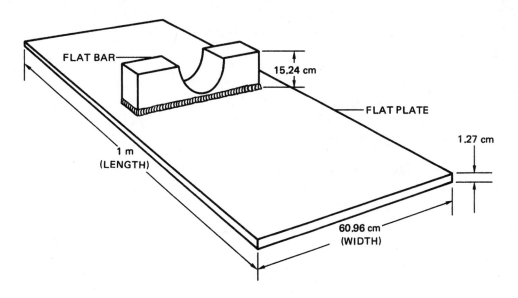

a. Find the overall height of the shaft support in centimeters.

a. _____

b. Express the length of the steel plate in millimeters.

b. _____

c. Express the width of the steel plate in centimeters.

c. _____

12.   Two steel angle samples are cut to the lengths shown. What is the total length of the two pieces in decimeters?   _____

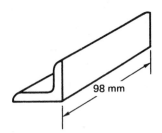 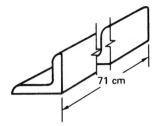

98 mm

71 cm

13.   This shaft is turned on a lathe from a piece of cold-rolled round stock.

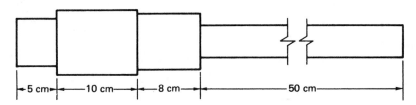

5 cm · 10 cm · 8 cm · 50 cm

a.   Find the total length in centimeters.   a. _____

b.   Find the total length in meters.   b. _____

14.   The length of standard wall pipe is 100 centimeters. Nine pieces of this pipe are welded together to form a continuous length. What is the length, in meters, of the welded section?   _____

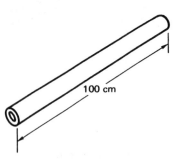

100 cm

Unit 22 EQUIVALENT UNITS

BASIC PRINCIPLES

Review the principles of equivalent units, and apply to these problems.
Review denominate numbers in Section I of the Appendix.
Study these tables of equivalent units.

ENGLISH LENGTH MEASURE

1 foot (ft)	=	12 inches (in)
1 yard (yd)	=	3 feet (ft)
1 mile (mi)	=	1,760 yards (yd)
1 mile (mi)	=	5,280 feet (ft)

METRIC LENGTH MEASURE

10 millimeters (mm)	=	1 centimeter (cm)
10 centimeters (cm)	=	1 decimeter (dm)
10 decimeters (dm)	=	1 meter (m)
10 meters (m)	=	1 dekameter (dam)
10 dekameters (dam)	=	1 hectometer (hm)
10 hectometers (hm)	=	1 kilometer (km)

Example: A length of steel angle is 12′ long. Express this measurement in inches.

1 foot (′) = 12 inches (″)
12 feet (′) x 12 inches (″) =
12′ x 12″ = 144″

PRACTICAL PROBLEMS

1. A piece of plate steel is shown.

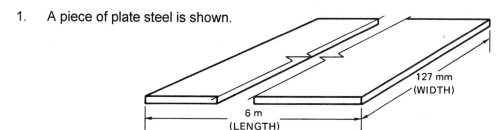

127 mm
(WIDTH)

6 m
(LENGTH)

a. Express the width in centimeters.

b. Express the length in centimeters.

a. _____

b. _____

2.   Express this 23′ length of the steel channel in inches.   _____

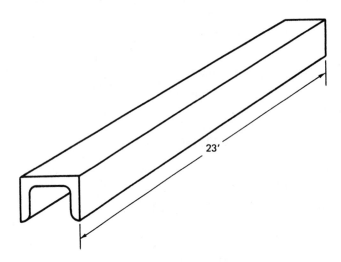

3.   Express the distance between hole centers in feet and a fractional part of a foot.   _____

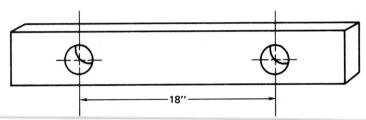

4.   Express this length of the round stock shown in feet.   _____

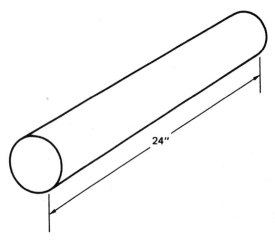

5.   Express the length of this pipe in inches.   _____

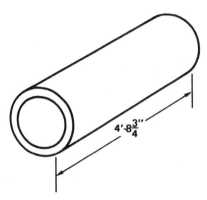

$4'-8\frac{3}{4}''$

6.   The fillet weld shown has 1 meter and 50 centimeters of weld in the joint.
       Express the total amount of weld in meters.   _____

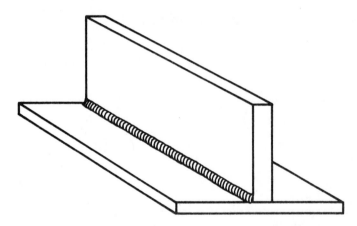

7. This tee-bracket is 18.625″ long. Express the measurement in feet, inches, and a fractional part of an inch.　　　　　————————

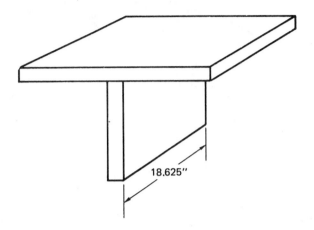

18.625″

8. A piece of I beam is 3′ 2.75″ long. Express this measurement in inches and a fractional part of an inch.　　　　　————————

9. A circular steel plate is flame cut and the center is removed.

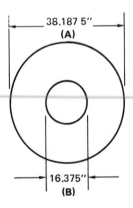

38.187 5″
(A)

16.375″
(B)

a. Express diameter A in feet, inches, and a fractional part of an inch.　　a. ————————

b. Express diameter B in feet, inches, and a fractional part of an inch.　　b. ————————

10. A flame-cut steel plate has the dimensions shown. Express each dimension in feet, inches, and a fractional part of an inch.

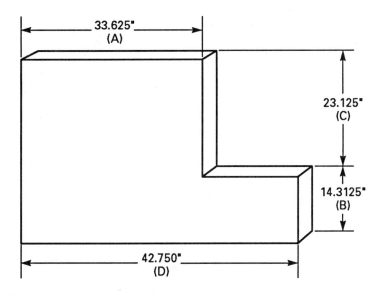

a. Dimension A

b. Dimension B

c. Dimension C

d. Dimension D

a. _____

b. _____

c. _____

d. _____

Unit 23 ENGLISH-METRIC EQUIVALENT UNITS

BASIC PRINCIPLES

Review and apply the principles of English-metric equivalent units, and apply them to these problems.

Review denominate numbers in Section I of the Appendix.

Study this table of English-metric equivalents.

ENGLISH–METRIC EQUIVALENTS

1 inch (in)	=	25.4 millimeters (mm)
1 inch (in)	=	2.54 centimeters (cm)
1 foot (ft)	=	0.3048 meter (m)
1 yard (yd)	=	0.9144 meter (m)
1 mile (mi)	≈	1.609 kilometers (km)
1 millimeter (mm)	≈	0.03937 inch (in)
1 centimeter (cm)	≈	0.39370 inch (in)
1 meter (m)	≈	3.28084 feet (ft)
1 meter (m)	≈	1.09361 yards (yd)
1 kilometer (km)	≈	0.62137 mile (mi)

When converting English to metric, or metric to English, use the table above.

Problem: Convert 2 meters into feet. 1 meter = 3.28084 feet.

```
3.28084   (feet)
x      2   (meters)
6.56168   (feet)
```

Problem: Convert 3 feet into meters. 1 foot = 0.3048 meters.

```
0.3048   (meters)
x      3   (feet)
0.9144   (meters)
```

PRACTICAL PROBLEMS

Note: Use this diagram for Problems 1 and 2.

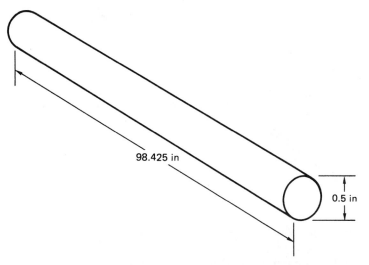

98.425 in

0.5 in

1.   The round stock is 98.425″ long. Express this length in meters. Round
     the answer to the nearest thousandth meter.   _____

2.   Find the diameter of the round stock to the nearest hundredth centimeter.   _____

3.   This I beam is 180 cm long and 3.2622 cm high. Round each answer to
     three decimal places.

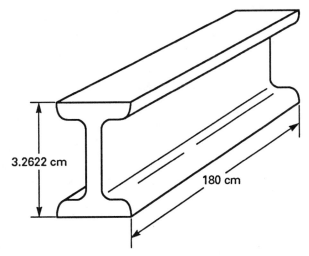

3.2622 cm

180 cm

     a.   Express the length in inches.   a. _____

     b.   Express the height in inches.   b. _____

4.   A piece of plate stock is shown.

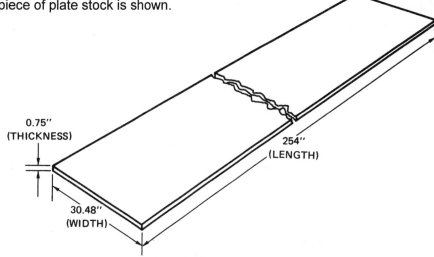

a.   Express the plate thickness to the nearest thousandth centimeter.   a. _____

b.   Express the plate width in centimeters. Round the answer to three
     decimal places.   b. _____

c.   Express the plate length in centimeters. Round the answer to two
     decimal places.   c. _____

5.   Express each dimension in centimeters. Round each answer to the
     nearest thousandth.

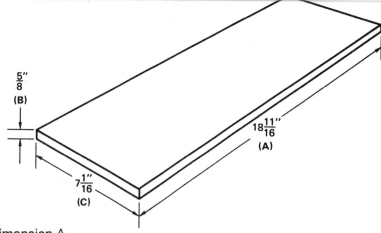

a.   Dimension A   a. _____

b.   Dimension B   b. _____

c.   Dimension C   c. _____

6.   Express each dimension in centimeters. Round each answer to the nearest thousandth.

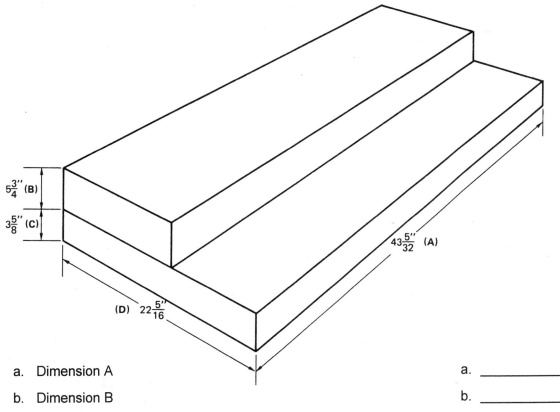

a.   Dimension A   a. _____

b.   Dimension B   b. _____

c.   Dimension C   c. _____

d.   Dimension D   d. _____

7.   Express each measurement in millimeters.

a.   ¹⁄₁₆ inch   a. _____

b.   ⅛ inch   b. _____

c.   ³⁄₁₆ inch   c. _____

d.   ¼ inch   d. _____

e.   ⅜ inch   e. _____

f.   ½ inch   f. _____

g.   ¾ inch   g. _____

h.   ¹¹⁄₁₆ inch   h. _____

Unit 24   COMBINED OPERATIONS WITH EQUIVALENT UNITS

BASIC PRINCIPLES

Review the principles of combined operations with equivalent units, and apply them to these problems.

Review the tables of equivalent units in Section II of the Appendix.

PRACTICAL PROBLEMS

Note: Use this diagram for Problems 1–3.

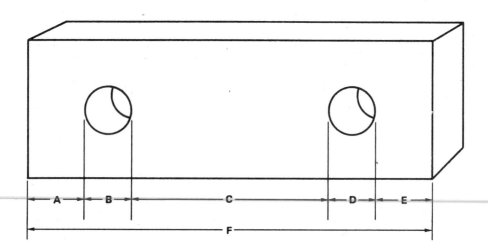

1.   This spacer block has these dimensions: A = 3.81 cm; B = 2.54 cm; C = 22.86 cm; D = 2.54 cm; E = 3.81 cm. Find dimension F to the nearest thousandth inch.

_____

2.   The dimensions on the spacer block are: F = 12 in; A = $1\frac{1}{4}$ in; B = $\frac{3}{4}$ in; D = $\frac{3}{4}$ in; E = $1\frac{1}{4}$ in. Find C in centimeters. Round the answer to two decimal places.

_____

3.   The dimensions on the spacer block are: E = 1.375 in; D = 0.875 in; A = 1.625 in; B = 0.875 in; C = 7.875 in. Find dimension F in millimeters. Round the answer to three decimal places.

_____

4.   This drawing shows a welded pipe support.

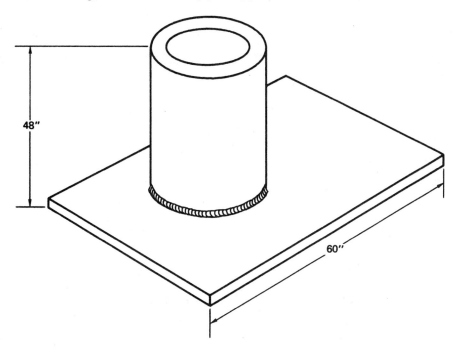

a. Express the height in meters.   a. _____

b. Express the width in meters.   b. _____

5.   This steel gusset is a right angle triangle. Round each answer to three
      decimal places.

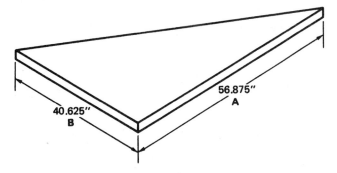

a. Express side A in centimeters.   a. _____

b. Express side B in centimeters.   b. _____

6.   A pipe bracket is shown. Round all answers to three decimal places.

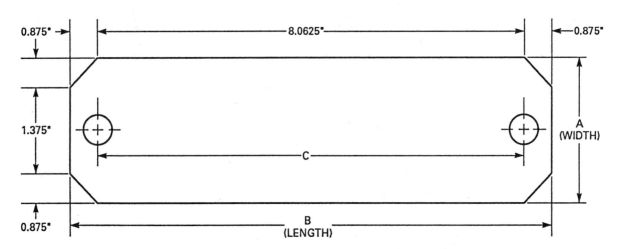

a. Find the width of the pipe bracket (dimension A) in millimeters.   a. _____

b. What is the length of the pipe bracket (dimension B) in millimeters?   b. _____

c. Find the distance between the center of the holes (dimension C) in
    centimeters.   c. _____

Note: Use this diagram for Problems 7 and 8.

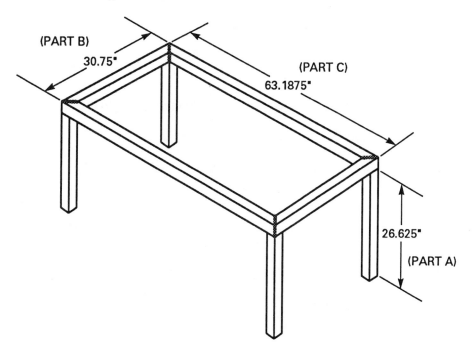

7.   A welder makes 20 of these table frames.

   a.   How many centimeters of square steel tubing are required to
        complete the order for Part A?   a. _____

   b.   How many centimeters of square steel tubing are required to
        complete the order for Part B?   b. _____

   c.   How many meters of square steel tubing are required to complete
        the order for Part C?   c. _____

8.   Each inch of square steel tubing weighs 0.1034 pounds. What is the
     total weight of all the tubing used for the 20 table frames? Express the
     answer to the nearest hundredth kilogram.   _____

Unit 25 PERIMETER OF SQUARES AND RECTANGLES

BASIC PRINCIPLES

Review the principles of perimeter of squares and rectangles, and apply them to these problems.

Review denominate numbers in Section I of the Appendix.

The distance around a figure is called the *perimeter.* The perimeter of a square is: side + side + side + side or $P = 4s$.

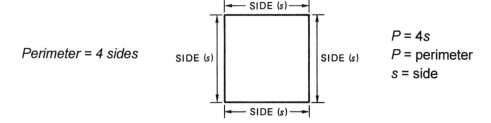

Perimeter = 4 sides

$P = 4s$
P = perimeter
s = side

Example: Find the perimeter of square A.

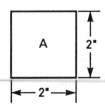

$2'' + 2'' + 2'' + 2'' = 8''$
or
$P = 4 \times 2'' = 8''$

Example: Find the perimeter of square B.

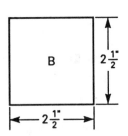

$2\frac{1}{2}'' + 2\frac{1}{2}'' + 2\frac{1}{2}'' + 2\frac{1}{2}'' = 10''$
or
$P = 4 \times 2\frac{1}{2}'' = 10''$

PRACTICAL PROBLEMS

Note: Use this information for Problems 1–6.

The measure of one side of each piece in a bundle of square plates is given. Find the perimeter of each piece.

1.   1¾ inches   _____

2.   3 inches   _____

3.   193.675 millimeters   _____

4.   14.5 inches   _____

5.   24.384 centimeters   _____

6.   0.6 meter   _____

Note: Use this information for Problem 7.

The perimeter of a rectangle is: length + width + length + width or $P = 2l + 2w$.

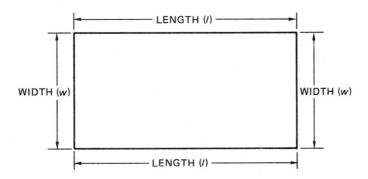

7.   What is the perimeter of this rectangle of steel?   _____

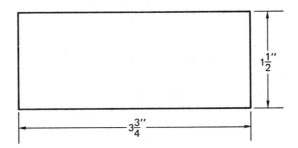

Unit 26 PERIMETER OF CIRCLES AND SEMICIRCULAR-SIDED FIGURES

BASIC PRINCIPLES

Review the principles of perimeter of circles and semicircular-sided figures, and apply them to these problems.

Review denominate numbers in Section I of the Appendix.

Definition: A circle is a plane curve all parts of which are equally distant from the center.

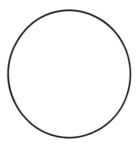

Circumference: The distance around the circle.

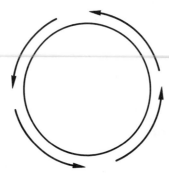

Radius: Straight line from the center to the curve equal to $\frac{1}{2}$ the diameter.

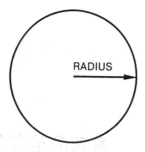

RADIUS

Diameter: A straight line through the center, ending at the curve, dividing the circle into two parts. Equal to twice the radius.

Note: It has been found by mathematical computation that the circumference of a circle is approximately 3.14 times the diameter of that circle. The number 3.14 is generally represented by the Greek letter pi (π). 3.14 is only an approximate number but for ordinary shop use is accurate enough.

The perimeter of a circle is called the *circumference.* To calculate the circumference (C) of a circle, multiply the diameter (D) times 3.14 (π or "pi").

The radius (r) multiplied by two and then multiplied by 3.14 (π) also equals the circumference.

Example: Using an automatic cutting torch, a circle is cut from ¼-inch steel plate. How many inches around the cut does the machine travel in completing the cutting operation? _____

$C = \pi D$

or

$C = 2\pi r$

where π = 3.14.

33″	= diameter of circle
D x pi	= circumference
33″ x 3.14	= 103.62″

33″	= diameter of circle
33″ ÷ 2	=radius
2) 33″	= 16½″ r
C	= 2 x 16½″ x 3.14
C	= 103.62″

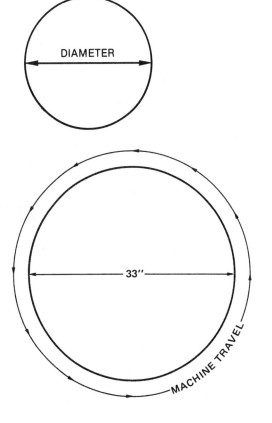

PRACTICAL PROBLEMS

1.   Circles A and B are cut from $\frac{3}{8}$-inch steel plate. How much more material
     is needed to form a lip around circle A than is needed for circle B?   _____

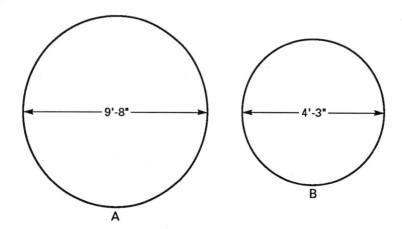

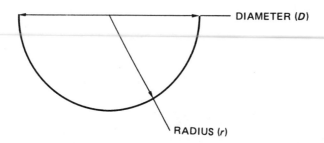

Hint: Find the circumference of A, then the circumference of B. Then subtract B from A.

Note: Use this information for Problems 2–11. The perimeter of a semicircle is:

$P = \pi r + D$ where $\pi = 3.14$.

2.   Find the circumference of the figure.   _____

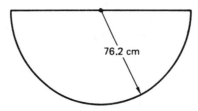

76.2 cm

Note: Use this information for Problems 3 and 4. The perimeter of a semicircular-sided form is:

$P = 2\pi r + 2l$ or $P = \pi D + 2l$
where $\pi = 3.14$.

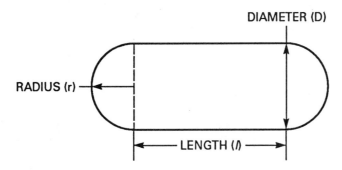

3. A semicircular-sided tank is welded in a shop. The bottom is cut from $\frac{1}{8}$-inch plate with the dimensions shown. How long is the piece of metal used to form the sides of the tank?   _____

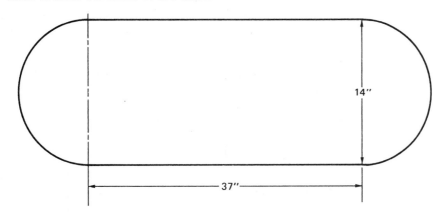

4. Find the distance around this semicircular-sided tank. Round the answer to four decimal places.   _____

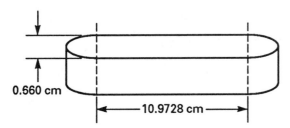

Angular Measure

Unit 27 ANGULAR MEASURE

BASIC PRINCIPLES

Circles are divided into degrees, minutes, and seconds.

The symbols used are: degrees (°), minutes (′), and seconds (″). The full circle equals 360 degrees; 1 degree = 60 minutes; 1 minute = 60 seconds; ¼ circle = 90 degrees; ½ circle = 180 degrees.

Examples:

> **Add:**
> 14° 31′ 14″
> + 2° 29′ 46″
> 16° 60′ 60″

Since 60″ = 1′, the one minute is carried to the minute column and added to the minutes. This gives us 16 degrees, 61 minutes, 0 seconds.

60′ = 1 degree; the one degree is carried to the degree column and added to the degrees.

Our answer is 17°, 1′, 0″.

> **Subtract::**
> 180° 19′ 14″
> - 90° 21′ 3″
> 89° 58′ 11″

3″ subtracted from 14″ = 11″

Since 21′ cannot be subtracted from 19′, we must borrow 1 degree from the degree column (1 degree = 60′), making 79′. This gives us 179 degrees, 79′, 14″.

> 179° 79′ 14″
> - 90° 21′ 3″
> 89° 58′ 11″

Multiply:

75° 36′ 19″

x   3

Multiply the entire problem: 3 x 19″, 36′, 75° = 225 degrees, 108 minutes, 57 seconds.

Beginning at the right, change the answer thus: 57″ is less than 1 minute, so it remains the same. 108 minutes contains one 60, so 1 minute is added to 225 degrees, leaving a remainder of 48 minutes.

Our answer is 226°, 48′, 57″.

PRACTICAL PROBLEMS

1.   93° 14′ 10″
  + 18° 59′ 58″

2.   45° 30′ 6″
  + 19° 14′ 0″

3.   180° 17′ 0″
  + 90° 45′ 19″

4.   90° 21′ 6″
  - 45° 19′ 12″

5.   75° 36′ 58″
  - 30° 14′ 26″

6.   283° 58′ 16″
  - 169° 9′ 43″

7.   16° 48′
  x   4

8.   45°
  x 8

9.   68° 41′ 35″
  x   2

Find the number of degrees in each of these parts of a circle. Note: A full circle equals 360 degrees.

10. $^1\!/_5$ _____

11. $^3\!/_4$ _____

12. $^1\!/_{16}$ _____

13. $^5\!/_8$ _____

Find the part of the circle represented by each angular measure.

14. 20°   _____

15. 40°   _____

16. 68°   _____

17. This pipe flange is drilled on a $4\frac{1}{2}$-inch radius and on a 5-inch radius. How many degrees farther apart are the holes in the 10-inch circle than the holes in the 9-inch circle?

_____

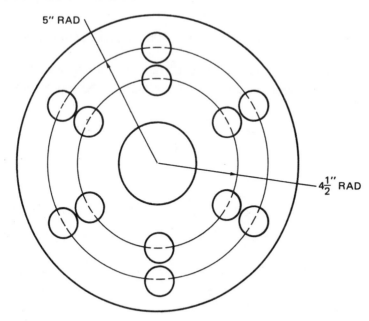

5″ RAD

$4\frac{1}{2}''$ RAD

18. How many degrees apart are the equally spaced holes in each of these pipe flanges?

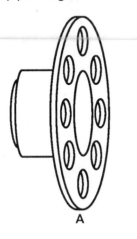

A

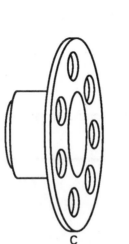

C

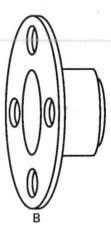

B

a. Flange A

b. Flange B

c. Flange C

a. _____

b. _____

c. _____

Unit 28   PROTRACTORS

BASIC PRINCIPLES

Review the principles of protractors, and apply them to these problems.

USING THE PROTRACTOR TO DRAW A 40 DEGREE ANGLE

Draw a straight line and place the protractor on the 0° to 180° bottom line.

Definition: An instrument in the form of a graduated semicircle used for drawing and measuring angles. (See illustration on Page 109.)

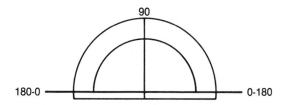

Make a mark on the center of the 0° to 180° line above the 90° mark on the protractor.

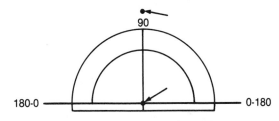

Make a mark on the 40° reading above the protractor.

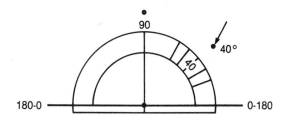

Draw a line from the mark on the 0° to 180° line through the mark on the 90° line; then draw a line from the base line mark through the 40° mark.

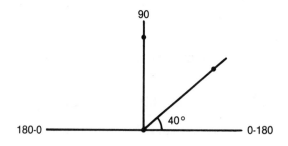

PRACTICAL PROBLEMS

1.   Using a protractor, measure each angle.

∠A = _____

∠B = _____

∠C = _____

∠D = _____

∠E = _____

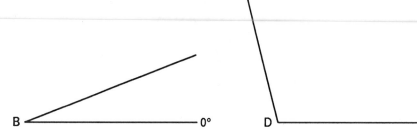

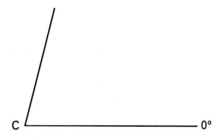

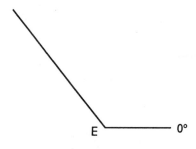

2.   Using a protractor, draw each angle.

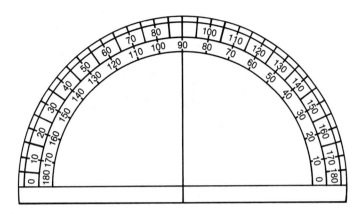

a.   45°   •————————————————

b.   30°   •————————————————

Computed Measure

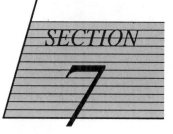

Unit 29 AREA OF SQUARES AND RECTANGLES

BASIC PRINCIPLES

Area is found by using the rule: side x side = area (always expressed as square measurement).

A piece of steel 3 feet x 4 feet equals 12 square feet in area, meaning that there are 12 1-foot squares in the piece of steel.

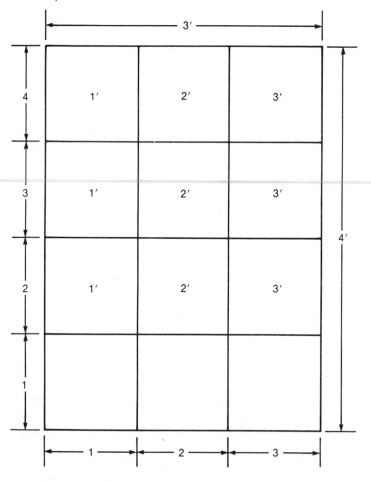

$$3' \times 4' = 12 \;\square\; ft$$

ENGLISH AREA MEASURE

1 square yard (sq yd)	= 9 square feet (sq ft)
1 square foot (sq ft)	= 144 square inches (sq in)
1 square mile (sq mi)	= 640 acres
1 acre	= 43,560 square feet (sq ft)

PRACTICAL PROBLEMS

Note: Use this information for Problems 1–4.

These squares are made from 16-gauge sheet metal. Find the area of each square.

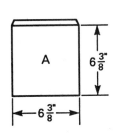

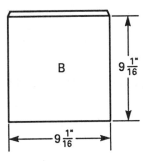

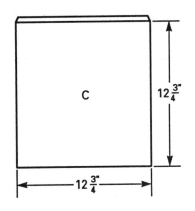

1.   Square A   _____

2.   Square B   _____

3.   Square C   _____

4.   Square D   _____

5.   Two hundred thirty-six pieces of sheet metal are cut to the specifications
     shown. What is the combined area of all of the pieces?   _____

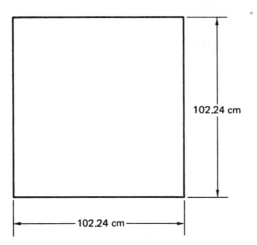

102.24 cm

102.24 cm

Note: Use this information for Problems 6–11.

The area of a rectangle is:

    area = length x width

        or

    $A = l \times w$

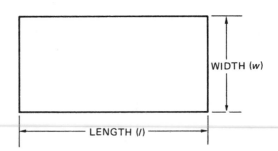

WIDTH (w)

LENGTH (l)

Note: Use this diagram for Problems 6 and 7.

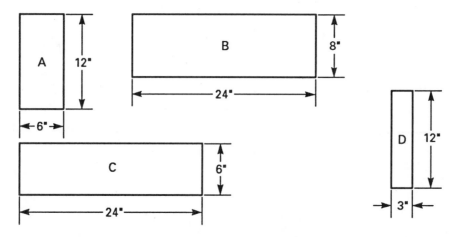

A   12"

6"

B   8"

24"

C   6"

24"

D   12"

3"

6. The four pieces of sheet metal are cut for a welding job.

    a. Find the area of rectangle A in square inches.

      a. _____

    b. Find the area of rectangle B in square inches.

      b. _____

    c. Find the area of rectangle C in square inches.

      c. _____

    d. Find the area of rectangle D in square inches.

      d. _____

    e. What is the total area of the pieces in square inches?

      e. _____

    f. Express the total area in square feet. Round the answer to two decimal places.

      f. _____

7. Which of the pieces has an area of one square foot?

      _____

8. A rectangular tank is made from plates with the dimensions shown. Find the total area of plate needed to complete the tank in square millimeters.

      _____

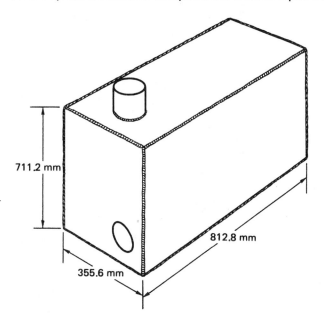

711.2 mm

812.8 mm

355.6 mm

9. How many square feet of material are needed for 20 of these tanks? Round the answer to two decimal places.

_____

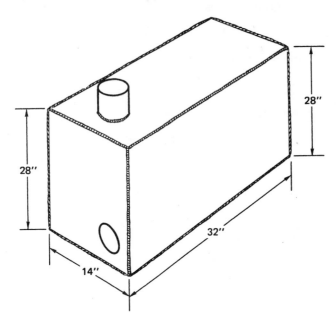

Note: Use this diagram for Problems 10 and 11.

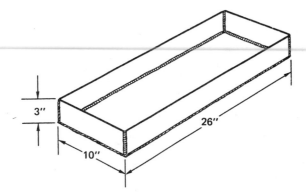

10. Five rectangular tool trays are welded as shown. Find, in square inches, the total amount of 16-gauge sheet metal needed for the job.

_____

11. A stock piece of 16-gauge sheet metal is 4 feet wide and 6 feet long. How many pieces are needed to complete the five tool trays?

_____

Unit 30 AREA OF TRIANGLES AND TRAPEZOIDS

BASIC PRINCIPLES

Review the principles of areas of triangles and trapezoids, and apply them to these problems.

Review denominate numbers in Section I of the Appendix.

Review the table of equivalent units of area measure in Section II of the Appendix.

Note: Use this information for Problems 1–6.

Rule: The area of a triangle is equal to $\frac{1}{2}$ of the base times the height.

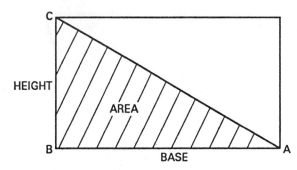

In a rectangle, the area is found by multiplying the base times the height (altitude). Drawing a diagonal from A to C divides the rectangle into two equal right triangles making the area of the triangle equal to $\frac{1}{2}$ of the rectangle. The formula for measuring the area of the triangle:

$$(\tfrac{1}{2} \text{ x base x height})$$

is true of all triangles.

Definition: A triangle is a three-sided figure containing three angles.

Rule: $\frac{1}{2}$ x [(*b*) base x (*h*) height]

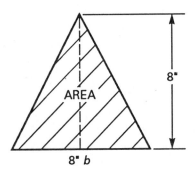

$\frac{1}{2}$ x (8″ X 8″) = $\frac{1}{2}$ x 64″ = 32 square inches (Area)

PRACTICAL PROBLEMS

Note: Use this information for Problems 1–4.

These four triangular shapes are cut from sheet metal. What is the area of each piece in square inches?

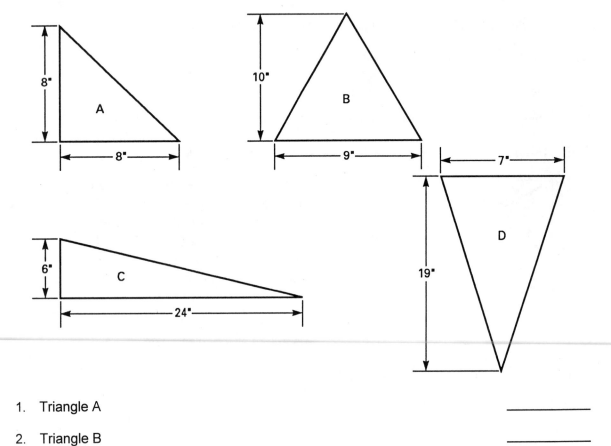

1.   Triangle A   _____

2.   Triangle B   _____

3.   Triangle C   _____

4.   Triangle D   _____

Note: Use this information for Problems 5 and 6.

Two pieces of sheet metal are cut into triangular shapes.

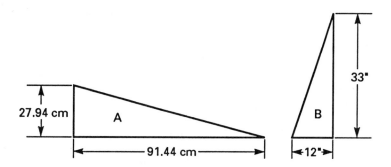

5. Find, in square centimeters, the area of triangle A.   _____

6. Find, in square inches, the area of triangle B.   _____

Definition: A trapezoid is a four-sided figure in which only two of the sides are parallel.

Note: Use this information for Problems 7–12.

The area of a trapezoid is:

area = $\frac{1}{2}$ x (longer base + shorter base) x height

$A = \frac{1}{2} \times (B + b) \times h$ or $A = \frac{B+b}{2} \times h$

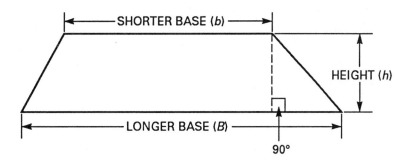

Note: Use this information for Problems 7–10.

These four support gussets are cut from ¼-inch plate. What is the area of each piece in square inches?

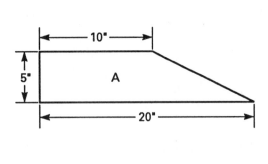

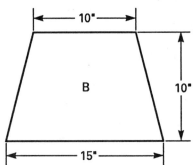

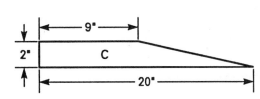

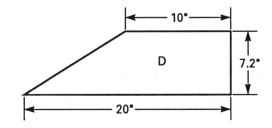

7.   Gusset A   _____

8.   Gusset B   _____

9.   Gusset C   _____

10. Gusset D   _____

11. One hundred twenty support gussets are cut as shown. Find, in square
     feet, the total area of steel plate needed for the complete order.   _____

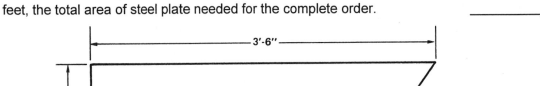

12. A welded steel bin is made from plates with these dimensions. Find, in square centimeters, the amount of plate needed to complete the bin.   _____

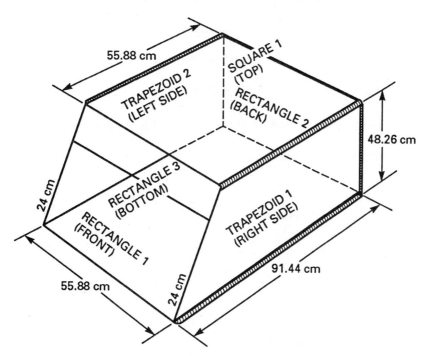

Hint: The total area equals the sum of the areas of the rectangles, the trapezoids, and the square.

Unit 31 AREA OF CIRCULAR FIGURES

BASIC PRINCIPLES

Review the principles of area of circular figures, and apply them to these problems.

Review denominate numbers in Section I of the Appendix.

Review the table of equivalent units of area measure in Section II of the Appendix.

Note: Use this information for Problems 1–4.

The area of a circle is:

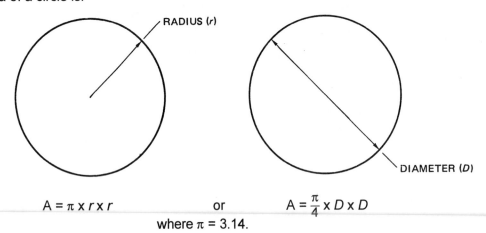

$$A = \pi \times r \times r \qquad \text{or} \qquad A = \frac{\pi}{4} \times D \times D$$

where π = 3.14.

Rule: The area of this semicircular-sided piece of steel is equal to the sum of the areas of the two semicircles and the rectangle.

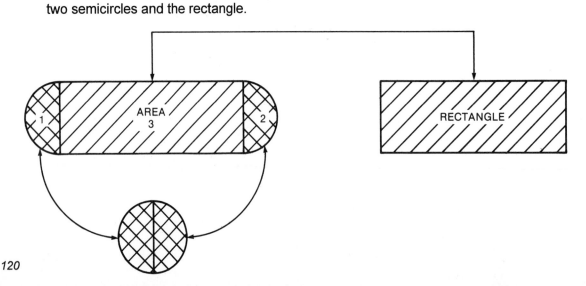

Two semicircles make a complete circle.

Area 1 + Area 2 + Area 3 = Area of the semicircle.

Rule: Pi (π) x (r x r) = Area of the circle.

W (width) x L (length) = Area of rectangle.

The area of a semicircular-sided figure is:

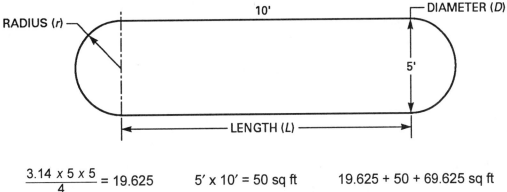

$$\frac{3.14 \times 5 \times 5}{4} = 19.625 \qquad 5' \times 10' = 50 \text{ sq ft} \qquad 19.625 + 50 + 69.625 \text{ sq ft}$$

$$A = (\pi \times r \times r) + (D \times L)$$
where π = 3.14.

PRACTICAL PROBLEMS

1. A steel tank is welded as shown. Find, in square inches, the area of the circular steel bottom.   _____

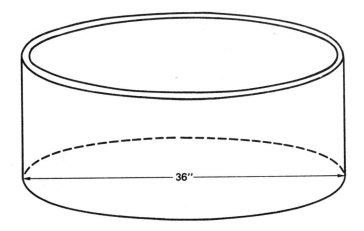

2. This tank bottom is cut from 0.635-cm steel plate. Express each area to the nearer thousandth square centimeter.

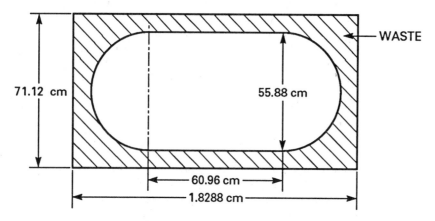

a. Find the area of the original plate.

b. Find the area of the tank bottom.

c. Find the area of the plate which is wasted.

a. _____

b. _____

c. _____

3. This tank bottom is cut from $^3/_{16}''$ steel plate. Express each answer in square inches.

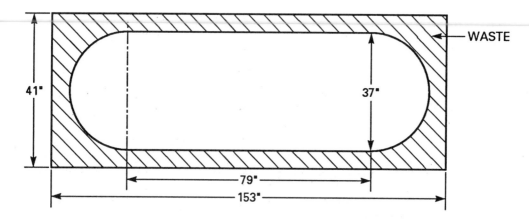

a. Find the area of the original plate.

b. Find the area of the semicircular-sided tank bottom.

c. Find the waste from the original plate.

a. _____

b. _____

c. _____

Note: Use this information for Problems 4–6.

Find the area of each semicircular-sided tank bottom. Express each area in square inches.

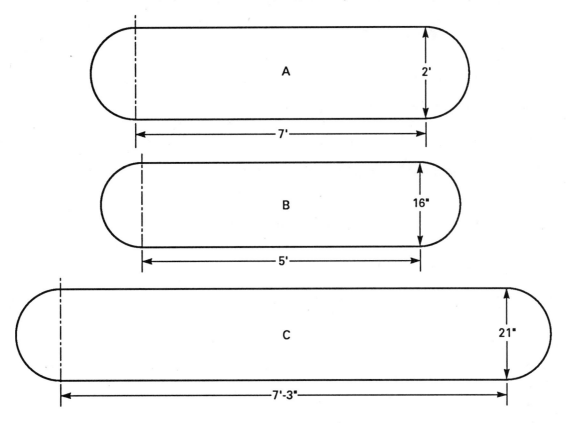

4. Bottom A   _____

5. Bottom B   _____

6. Bottom C   _____

Unit 32   VOLUME OF CUBES AND
RECTANGULAR SOLIDS

BASIC PRINCIPLES

Review the principles of volume of cubes and rectangular solids, and apply them to these problems.

Review denominate numbers in Section I of the Appendix.

Study this table of equivalent units of volume measure for solids.

ENGLISH VOLUME MEASURE FOR SOLIDS

| 1 cubic yard (cu yd) | = 27 cubic feet (cu ft) |
| 1 cubic foot (cu ft) | = 1,728 cubic inches (cu in) |

Note: Use this information for Problems 1–5.

The amount of space occupied in a three-dimensional figure is called the *volume.* Volume is also the number of cubic units equal in measure to the space in a solid.

The volume of a cube is:

volume = side x side x side

or

$V = s \times s \times s$

Definition: A cube is a solid with six equal square sides.

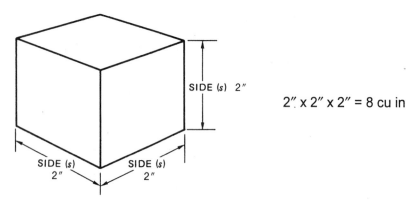

SIDE (s) 2″

$2″ \times 2″ \times 2″ = 8$ cu in

SIDE (s)
2″

SIDE (s)
2″

Note: See Appendix for explanation.

PRACTICAL PROBLEMS

1. A solid cube of steel is cut to these dimensions. Find the volume of the cube in cubic inches.   _____

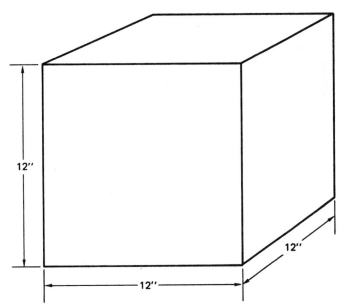

2. Five pieces of 5.08-cm solid square bar stock are cut to the specifications shown. Find the total volume of the pieces in cubic centimeters. Round the answer to three decimal places.   _____

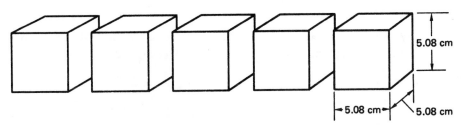

3. Two pieces of square stock are welded together. Find, in cubic feet, the total volume of the pieces. Round the answer to three decimal places. _____

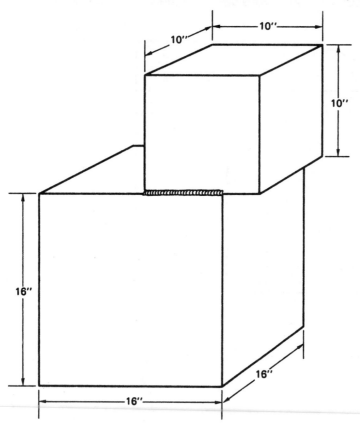

4. A spacer plate has five pieces of cold-rolled square bar welded to it. Find, in cubic inches, the total volume of the five blocks. Round the answer to three decimal places. _____

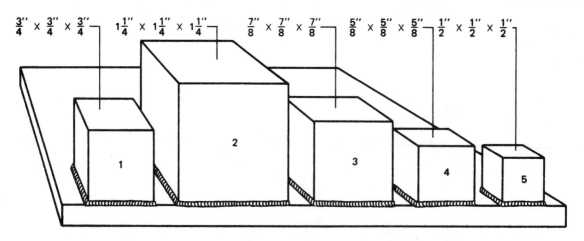

5. Bronze stock is melted and poured into a mold to form this cube. What is
    the volume of the completed cube in cubic inches? Round the answer to
    three decimal places.   _____

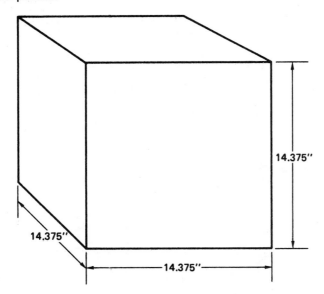

14.375"

14.375"

14.375"

Note: Use this information for Problems 6–12.

The volume of a rectangular solid is:

volume = length x width x height

or

$V = l \times w \times h$

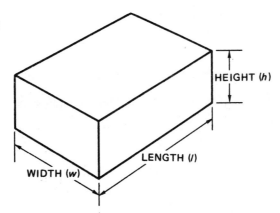

HEIGHT (h)

LENGTH (l)

WIDTH (w)

Note: Use this information for Problems 6–8.

Find, in cubic inches, the volume of each steel bar.

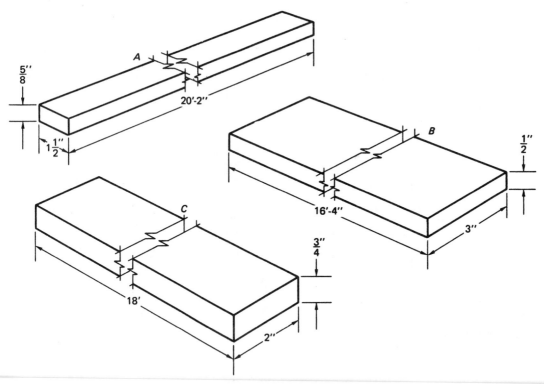

6.   Steel bar A.   _____

7.   Steel bar B.   _____

8.   Steel bar C.   _____

Find the volume of each rectangular solid. Round to three decimal places when needed.

9. l = 12 in; w = 8 in; h = 10 in   _____

10. l = 0.84 m; w = 0.46 m; h = 0.91 m   _____

11. l = 18.875 in; w = 11.125 in; h = 23.625 in   _____

12. l = 14$\frac{3}{4}$ in; w = 9$\frac{5}{8}$ in; h = 1$\frac{1}{4}$ in   _____

Unit 33 VOLUME OF CYLINDRICAL SOLIDS

BASIC PRINCIPLES

Review the principles of volume of cylindrical solids, and apply them to these problems.

Review denominate numbers in Section I of the Appendix.

Review the table of equivalent units of volume measure for solids in Section II of the Appendix.

Round all answers to three decimal places when needed.

Note: Use this information for Problems 1–10.

The volume of a cylindrical solid is:

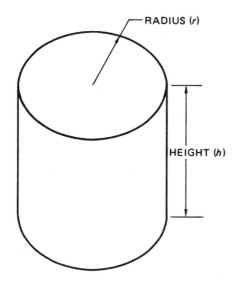

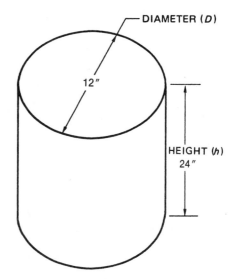

$$V = \pi \times r \times r \times h$$

$$
\begin{aligned}
V &= 3.14 \times 6'' \times 6'' \times 24'' \\
&= 3.14 \times 36'' \times 24'' \\
&= 113.04 \times 24'' \\
&= 2712.96 \text{ cu in}
\end{aligned}
$$

where $\pi = 3.14$.

PRACTICAL PROBLEMS

Find, in cubic inches, the volume of each piece of round stock.

1. $D = 10$ in; $h = 60$ in   _____

2. $D = 48$ in; $h = 48$ in   _____

3. $D = 8.125$ in; $h = 59.875$ in   _____

4. $D = 10.625$ in; $h = 72.75$ in   _____

Remember: All dimensions must be given in the same unit of measure before multiplying, that
is, inches by inches, feet by feet, and so on.

Find, in cubic feet, the volume of each cylinder.

5. $r = 12$ in; $h = 48$ in   _____

6. $r = 3$ in; $h = 120$ in   _____

7. $D = 12$ in; $h = 24$ in   _____

8. $D = 8.375$ in; $h = 22.125$ in   _____

9. Find, in cubic inches, the volume of 17 of these small welded hydraulic
    tanks.   _____

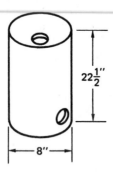

$22\frac{1}{2}''$

$8''$

Rule: The volume of a semicircular-sided tank is equal to the area of the base times the height.

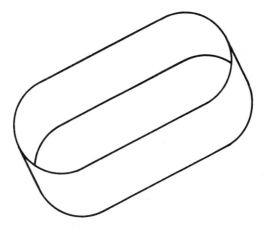

Example: The bottom of the tank is a rectangle and two semicircles:

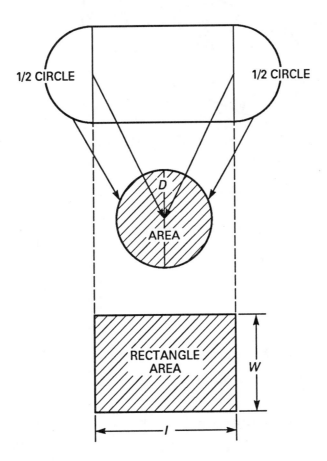

Therefore, to find the area we must find the area of a circle and the area of a rectangle.

Area of a circle = πr^2 (3.14 x *r* x *r*)

Area of a rectangle = *l* x *w* (length x width)

Then, by following the rule (the volume of a semicircular-sided tank is equal to the area of the base times the height), we have:

$$[(\pi \times r^2) + (l \times w)] \times (h) = \text{Volume}$$

Example: A semicircular tank has the following dimensions:

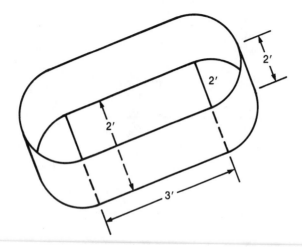

Find the volume:

Find the area of a two-foot diameter circle:

2″ x 12″ = 24″

Thus,

D = 24″, r = 12″

Then, $A = \pi r^2$ (*r* x *r*) = r^2

A = 3.14 x 12″ x 12″ = 452.16 square inches

Find the area of the rectangle:

3′ x 2′ = 36″ x 24″ = 864 square inches

Add the areas:

452.16 square inches
+ 864.00 square inches
1,316.16 square inches (discard remainder less than 5)

Multiply by height:

2′ = 24″

Thus,

1,316   square inches
x   24
5264
2632
31,584   cubic inches = Volume

10. What is the volume of this semicircular-sided solid in cubic feet? _____

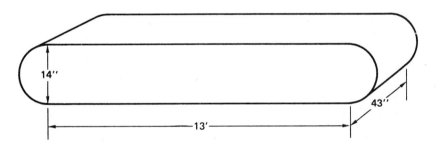

11. Two semicircular-sided tanks are shown. The dimensions of one tank are exactly twice the dimensions of the other tank. Is the volume of the larger tank twice the volume of the smaller tank? Explain. _____

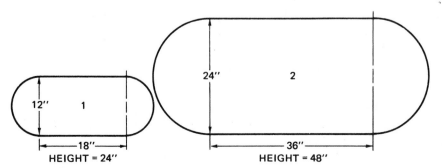

Unit 34 VOLUME OF RECTANGULAR CONTAINERS

BASIC PRINCIPLES

Review the principles of volume of rectangular containers, and apply them to these problems.

Review denominate numbers in Section I of the Appendix.

Review the tables of equivalent units of volume measure for solids in Section II of the Appendix.

Review the formulas for volumes in Section III of the Appendix.

Study these tables of volume measure equivalents.

ENGLISH VOLUME MEASURE EQUIVALENTS

1 gallon (gal)	=	0.133681 cubic foot (cu ft)
1 gallon (gal)	=	231 cubic inches (cu in)

METRIC VOLUME MEASURE EQUIVALENTS

1 cubic decimeter (dm^3)	=	1 liter (L)
1,000 cubic centimeters (cm^3)	=	1 liter (L)
1 cubic centimeter (cm^3)	=	1 milliliter (mL)

Note: Use inside dimensions when finding the volume of containers. If these dimensions are not given, the wall thickness is subtracted from the outside dimensions.

Volume in gallons = l'' x w'' x h'' divided by 231 cu in $V = \dfrac{l \times w \times h}{231}$

PRACTICAL PROBLEMS

Find the volume, in gallons, of each rectangular welded tank. These are inside dimensions.

Round each answer to three decimal places.

1. l = 9.875 in; w 6.1875 in; h = 24.125 in   _____

2. l = $12\frac{3}{4}$ in; w $14\frac{7}{8}$ in; h = $36\frac{1}{4}$ in   _____

3. l = 36 in; w = 18 in; h = 48 in   _____

4.   *l* = 23.5 in; *w* = 23.5 in; *h* = 34.5 in _____

5.   The dimensions on this box are inside dimensions. Find the number of cubic inches of volume in the box. _____

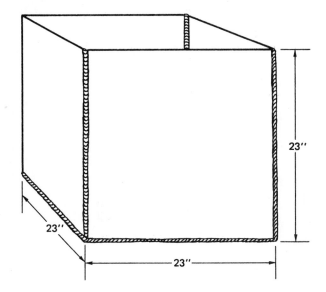

The dimensions on this welded square box are inside dimensions. Find the number of liters that the tank can hold. Round the answer to three decimal places. _____

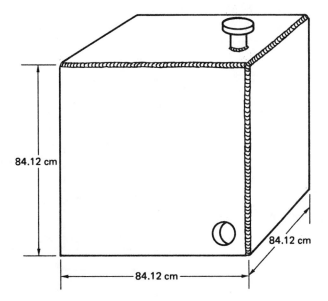

Note: Use this information for Problems 7 and 8.

Welded tanks A and B are made from $\frac{1}{8}$-inch steel plate.

$\frac{1}{8}''$ thickness must be deducted from the given measurement twice to give correct volume inside.

$$10\frac{5}{8}'' - (\frac{1}{8}\text{Æ} + \frac{1}{8}'') + 10\frac{3}{8}''$$

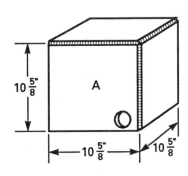

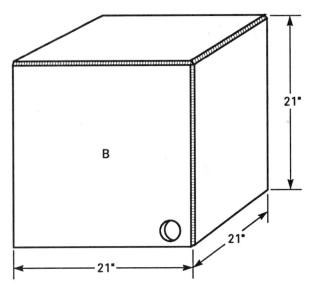

7. Find the volume of tank A. Round the answer to the nearest thousandth cubic inch.

    _____

8. Find the volume of tank B.

    _____

9. The dimensions of welded storage tanks C and D are inside dimensions. The dimensions of tank D are exactly twice those of tank C. Is the volume of tank D twice the capacity of tank C?

    _____

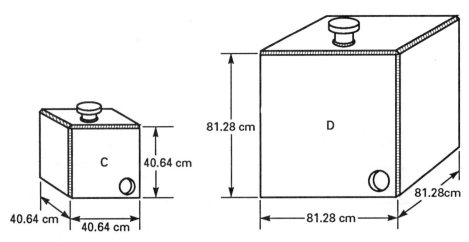

10. Square tanks A, B, C, and D are welded and filled with a liquid. Which of the tanks has a volume closest to one gallon? The dimensions are inside dimensions.

_____

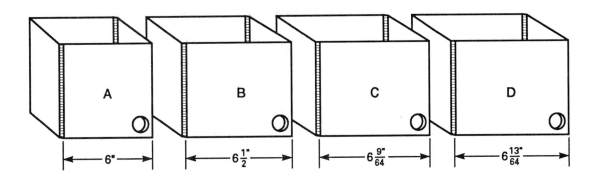

11. Nine fuel storage tanks for pickup trucks are welded. The dimensions are inside dimensions.

_____

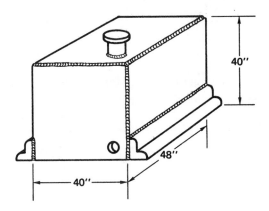

a. What is the total volume, in cubic inches, of the entire order of tanks?

a. _____

b. What is the total volume in cubic feet?

b. _____

12. A rectangular tank is welded from ⅛-inch steel plate to fit the specifications shown. How many gallons does the tank hold? Round the answer to three decimal places. The dimensions are inside dimensions. _____

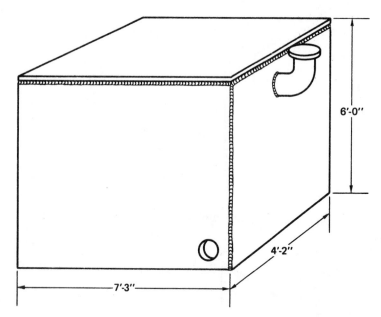

6'-0''

4'-2''

7'-3''

13. This welded tank has two inside dimensions given. The tank holds 80.53 gallons of liquid. Find, to the nearest tenth inch, dimension x. _____

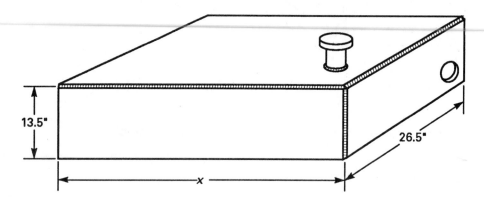

13.5"

26.5"

x

14. This rectangular welded tank is increased in length, so that the volume, in gallons, is doubled. What is the new length (dimension *x*) after the welding is completed?   _____

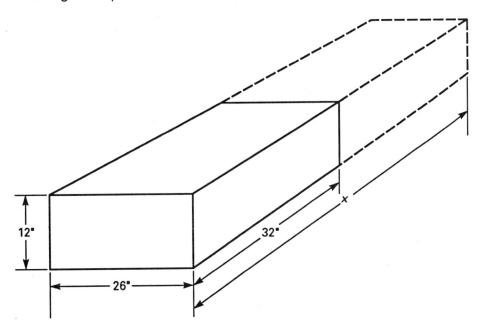

15. A pickup truck tank holds 89 liters of gasoline. Two auxiliary tanks are constructed to fit into spaces under the fenders of the truck. What is the total volume of the two tanks plus the original tank to the nearest thousandth liter?   _____

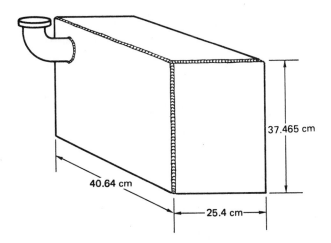

16. This welded steel tank is damaged. The section indicated is removed and a new bulkhead welded in its place. How many fewer liters will the tank hold after the repair? Round the answer to three decimal places. _____

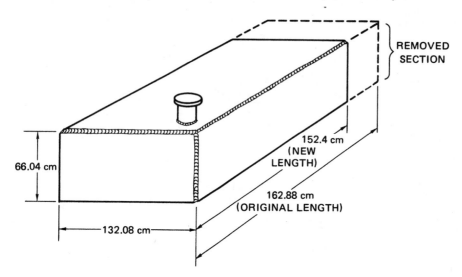

REMOVED SECTION

152.4 cm
(NEW LENGTH)

66.04 cm

162.88 cm
(ORIGINAL LENGTH)

132.08 cm

Unit 35   VOLUME OF CYLINDRICAL AND COMPLEX CONTAINERS

BASIC PRINCIPLES

Review the principles of volume of cylindrical and complex containers, and apply them to these problems.

Review denominate numbers in Section I of the Appendix.

Review the tables of equivalent units of volume measure in Section II of the Appendix.

Use pi (π) = 3.14.

Volume = $\pi r^2 h$ or $\frac{\pi}{4} D^2 h$

Example: A pipe 10″ long has an inside diameter of 4″. Find the volume.

V = Pi x radius squared x height, or
V = Pi divided by 4 x D squared x height.
Thus, V = 3.14 x 2″ x 2″ x 10″, or
V = 3.14 x 4″ x 10″
Then, V = 12.561 x 10″
V = 125.6 cu in

PRACTICAL PROBLEMS

1. A pipe with an inside diameter of 10 inches is cut into three pieces. Find the volume of each piece, in cubic inches.

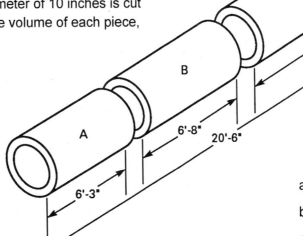

a. Piece A

b. Piece B

c. Piece C

a. _____

b. _____

c. _____

2. An outside storage tank is welded. The dimensions given are inside dimensions.

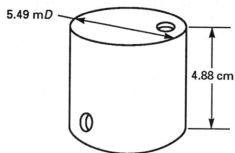

5.49 m*D*

4.88 cm

a. Find, in cubic meters, the volume of the tank. Round the answer to five decimal places.

a. _____

b. Find, in liters, the volume of the tank. Round the answer to two decimal places.

b. _____

3. The dimensions on these three cylindrical welded tanks and connecting pipes are inside dimensions. The tanks are connected as shown and are filled with liquid. The system is completely filled, including the connecting pipes. What is the total volume of the system? Round the answer to the nearest thousandth gallon.

_____

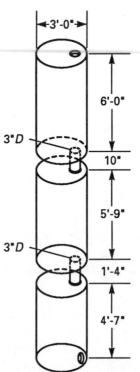

3'-0"

6'-0"

3"*D*

10"

5'-9"

3"*D*

1'-4"

4'-7"

4. A welded manifold is constructed as shown. Find, in cubic inches, the volume of the system (I.D. = inside diameter). _____

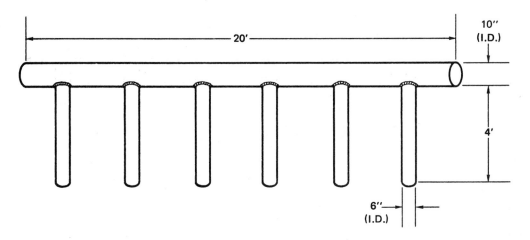

5. A 90° two-piece elbow is cut and welded from a 60.96-cm inside diameter pipe. Find the volume of the elbow to the nearest thousandth cubic meter. _____

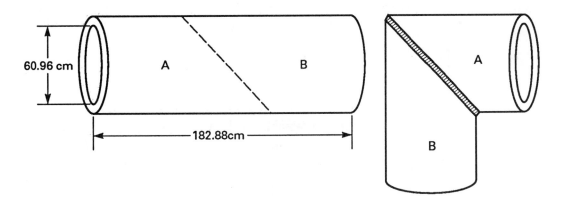

6. A length of welded irrigation pipe has the dimensions shown. Twenty of these lengths are welded together. What is the total volume of the welded pipes? Round the answer to the nearest thousandth gallon.   _____

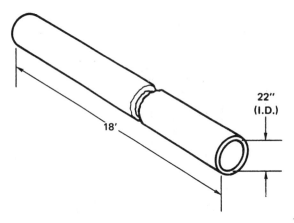

7. A weldment consisting of a semicircular-sided tank and a steel angle frame is constructed as shown. The dimensions given are inside dimensions. What is the volume of the complete tank to the nearest gallon?   _____

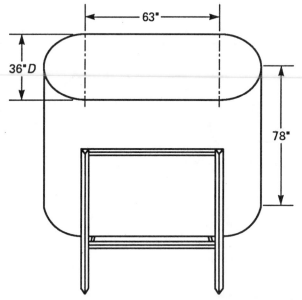

8. Using semicircular-sided pipes, this manifold system is welded. The dimensions given are inside dimensions.

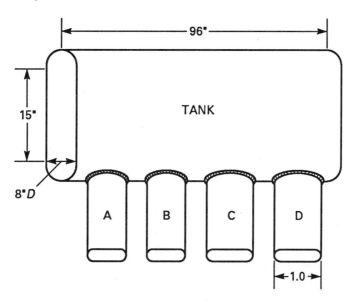

PIPE A = 2″ *D*; 4″ I.D.; 14″ HEIGHT

PIPE B = 2″ *D*; 4″ I.D.; 14″ HEIGHT

PIPE C = 3″ *D*; 5″ I.D.; 14″ HEIGHT

PIPE D = 3″ *D*; 7″ I.D.; 14″ HEIGHT

a. Find, in cubic inches, the total volume of the pipes.

b. Find, in cubic inches, the volume of the tank.

c. Find, in gallons, the volume of the entire manifold system. Round the answer to three decimal places.

a. _____

b. _____

c. _____

9. Two settling tanks are welded together. The dimensions given are inside dimensions. Find, in gallons, the volume of the entire system. Round the answer to the nearest thousandth gallon. _____

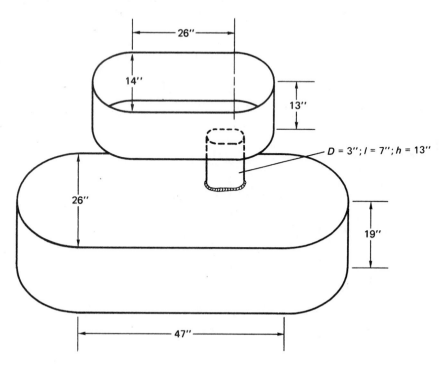

Unit 36 MASS (WEIGHT) MEASURE

BASIC PRINCIPLES

Review the principles of mass and weight measure, and apply them to these problems.

Review denominate numbers in Section I of the Appendix.

The mass (weight) of one cubic inch of steel is 0.2835 pound. The mass (weight) of one cubic centimeter of steel is 7.849 grams.

Example: A piece of 9″ round stock has a diameter of 4″.

Calculate the weight of the round stock.

Hint: Find the cubic inches in the piece of steel.

V = radius squared x pi x height.

V = 2″ x 2″ x 3.14 x 9″

V = 4″ x 3.14 x 9″

V = 12.56 sq in x 9″

V = 113.04 cu in

Weight = 113.04 cu in x 0.2835 (weight of 1 cu in of steel)

Weight = 32.036 pounds

When an object is pushed, pulled, or lifted, the *mass* (*quantity*) of the object is being moved; when put on a scale, the gravitational *pull* (*weight*) is measured. The mass and the volume sometimes form equivalent measures. When this is true, the measurements are considered as constants. For instance, 7.489 grams of steel equals 1 centimeter cubed (centimeter x centimeter x centimeter).

PRACTICAL PROBLEMS

Round the answer to four decimal places when needed.

1. Fourteen pieces of cold-rolled steel shafting are sawed as shown. What is the total weight of the 14 pieces of steel in pounds? _____

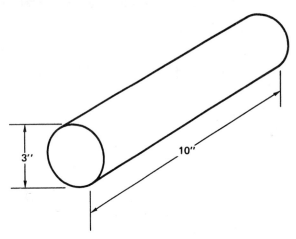

2. A welder flame cuts 10 roof columns from 8″ pipe as shown. Find the total weight of the columns (O.D. = outside diameter). _____

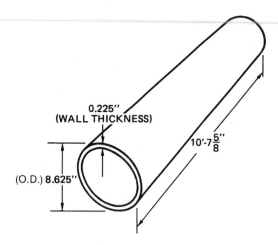

3. Steel angle legs for a tank stand have the dimensions shown. Find the
    weight of twenty legs in kilograms.

    _____

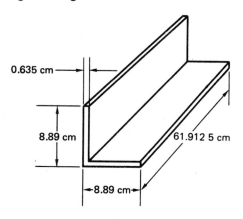

0.635 cm

8.89 cm

8.89 cm

61.912 5 cm

4. A circular tank bottom is cut as shown.

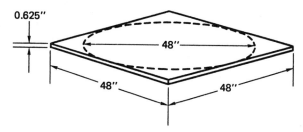

0.625"

48"

48"

48"

a. Find the weight of the circular bottom.

a. _____

b. Find the weight of the wasted material.

b. _____

5. An open-top welded bin is made from ¼-inch plate steel. What is the
    total weight, in pounds, of the five pieces used for the bin?

    _____

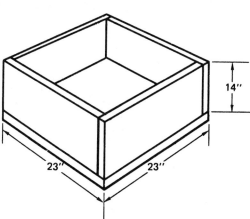

14"

23"

23"

6.   Sixteen circular blanks for sprockets are cut from 1.27-cm plate.

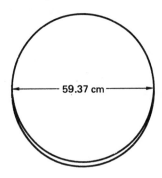

59.37 cm

a.   What is the weight of one blank in kilograms?   a. _____

b.   What is the weight of all of the blanks in kilograms?   b. _____

7.   Pieces of $\frac{3}{8}$-inch bar stock are used for welding tests. Find, in pounds,
     the weight of one piece of the bar stock.   _____

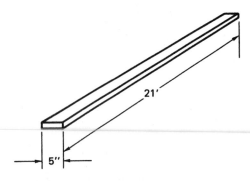

21′

5″

8. A column support gusset is shown. Find, in pounds, the weight of 52 of
    these gussets.   _____

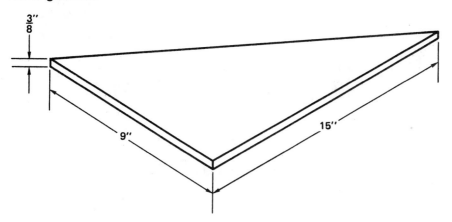

9. Find, in pounds, the weight of this adjustment bracket.   _____

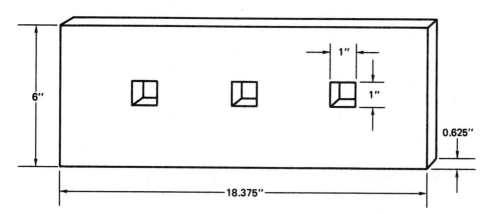

Pieces and Lengths

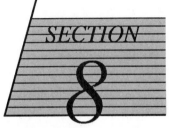

Unit 37 STRETCHOUTS OF SQUARE AND RECTANGULAR SHAPES

BASIC PRINCIPLES

Study the principles of stretchouts of square and rectangular shapes, and apply them to these problems.

When metal is bent into corners, the length of flat material needed to make the corner depends on the bend angle, the bend radius, and the material thickness. The bend allowance is usually found from a chart, but this formula gives an approximate answer for right angles. The formula for 90° minimum radius bends is:

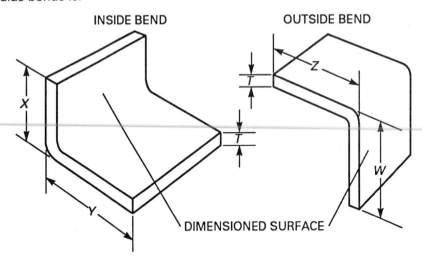

INSIDE BEND

OUTSIDE BEND

DIMENSIONED SURFACE

$$L = X + Y + \frac{1}{2}T \qquad\qquad L = Z + W - \frac{1}{2}T$$

where

L = length of flat material before bending

T = thickness of the material

Add $\frac{1}{2}$ the metal thickness for each inside bend and subtract $\frac{1}{2}$ the metal thickness for each outside bend.

A bend is defined as inside when the inside of the formed corner is seen by looking at the dimensioned surface (usually the outside surface).

Example: Assume that all welded corners are cut so that one piece overlaps the end of the other. Express all answers in the form length x width and round to three decimal places when needed.

This pipe is made from $\frac{1}{8}$-inch sheet metal. Three 90° outside bends are required to prepare this square for welding. Find the size of sheet metal needed to bend and weld the pipe. Express the answer in feet and inches.

Known: T = thickness ($\frac{1}{8}''$ or 0.125″)
Z = leg size (18.375″)
L = length (8′-9.875″)
3 outside bends required

And, $\frac{1}{2}T = \frac{1}{16}''$ or 0.0625″

Then: 18.250″ (Z) - 0.0625″ ($\frac{1}{2}T$) = 18.1875″
18.375″ (Z) - 0.0625″ ($\frac{1}{2}T$) = 18.3125″
18.375″ (Z) - 0.0625″ ($\frac{1}{2}T$) = 18.3125″
18.250″   = 18.250″

Added: 18.1875 in
18.3125 in
18.3125 in
+ 18.250 in
73.0625 in

Which is 73″ + 0.0625″ or 6′ 1$\frac{1}{16}$″

Then: $73 \times \dfrac{1}{12}$ = $\dfrac{73}{12}$

= $6\dfrac{1}{12}$

= 6′ 1″

Answer: 6′-1.0625″ x 8′-9.875″

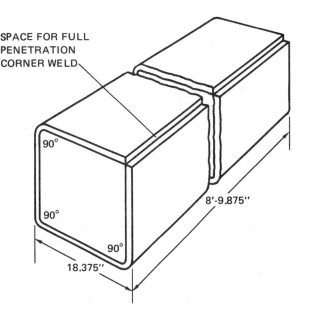

SPACE FOR FULL
PENETRATION
CORNER WELD

90°

90°

90°

18.375″

8′-9.875″

PRACTICAL PROBLEMS

1. This machinery cover is of a square-welded, two-piece, 0.64 cm steel plate design. Two 90° outside bends are required for preparation. Find the size of each piece used for the weldment.

_____

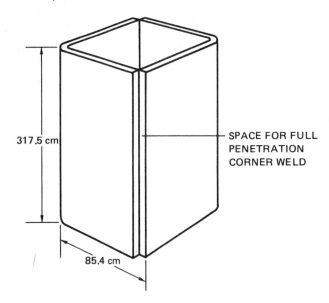

317.5 cm

SPACE FOR FULL
PENETRATION
CORNER WELD

85.4 cm

2. Two outside 90° bends are used to shape a section of transfer gutter. Thirteen sections are welded together to complete the run. Find, in square inches, the amount of $^1\!/_{16}''$ steel needed to complete the total of 13 sections. Round the answer to four decimal places.

_____

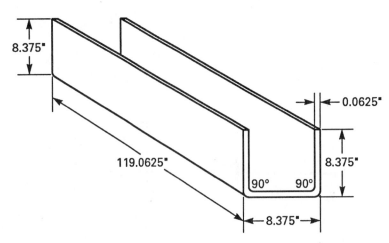

8.375"

0.0625"

119.0625"

8.375"

90°   90°

8.375"

3.   Four inside 90° bends are used in this gasoline tank. Find the size of $\frac{1}{8}''$ plate steel used to complete the tank body. Make no allowance for a weld gap.

_____

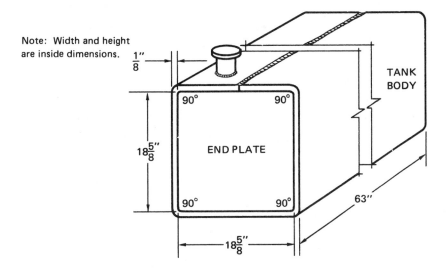

Note: Width and height are inside dimensions.

$\frac{1}{8}''$

90°   90°

$18\frac{5}{8}''$ END PLATE

90°   90°

$18\frac{5}{8}''$

TANK BODY

63''

4.   Forty-three open-end welded steel storage boxes are welded from 0.48 cm plate steel. The plate thickness is enlarged to show detail. Find the size of plate used for one storage box.

_____

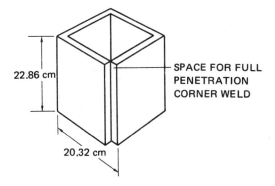

22.86 cm

SPACE FOR FULL PENETRATION CORNER WELD

20.32 cm

5. This parts bin is 4″ square and 8″ long. The bins are made from 4″ wide strips. Find the number of bins (without bottoms) that can be fabricated from the sheet of metal shown. Disregard cutting losses and bend allowances.

_____

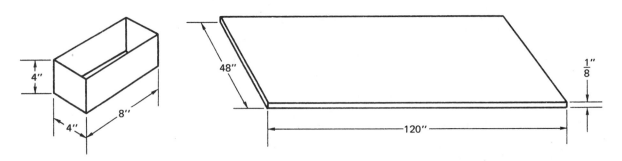

6. Four 90° outside bends are required for the tank shown. Find the size, in feet, of the steel piece needed to bend the main body of the tank. The material thickness is $\frac{1}{8}$″. Make no allowance for a weld gap.

_____

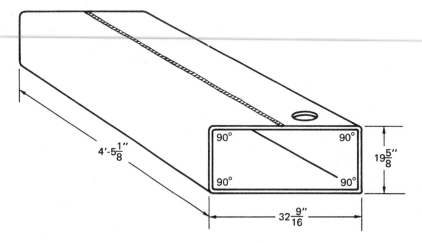

7.   A welded high-pressure hydraulic tank is shown. Find the size of the
     0.635 cm sheet of steel plate used to construct the tank. Outside bends
     are used.   _____

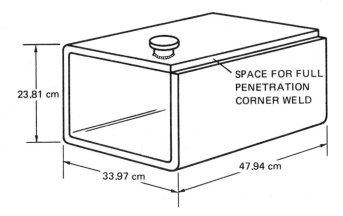

SPACE FOR FULL
PENETRATION
CORNER WELD

23.81 cm

33.97 cm

47.94 cm

Unit 38   STRETCHOUTS OF CIRCULAR AND SEMICIRCULAR SHAPES

BASIC PRINCIPLES

Study the principles of stretchouts of circular and semicircular shapes, and apply them to these problems.

When a cylindrical (or semicircular) shape is stretched out, the length is equal to the circumference (C) of the cylinder (or perimeter of the semicircular-sided shape), using the average of the inside and outside diameters. The average diameter is found by subtracting a wall thickness from the outside diameter (O.D.) or by adding a wall thickness to the inside diameter (I.D.).

Use (pi) π = 3.14. Express all answers in the form length x width and round to three decimal places when needed.

Example: Find the size of the $\frac{1}{8}$" metal needed to form this circular drainpipe. Express the answer in inches.

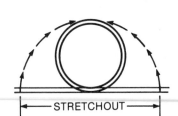

Average diameter (D) = $18\frac{5}{8}$" - $\frac{1}{8}$" = $18\frac{4}{8}$" or $18\frac{1}{2}$"

$18\frac{1}{2}$" = 18.5"

Thus,   C = pi (3.14) x D
              C = 3.14 x 18.5"
              C = 58.09"

Now, length = 6' $5\frac{3}{8}$" = 6' + $5\frac{3}{8}$" = 72" + $5\frac{3}{8}$" = $77\frac{3}{8}$"
                    $77\frac{3}{8}$" = 77.375"

Answer: 58.09" (C) x 77.375" (l)

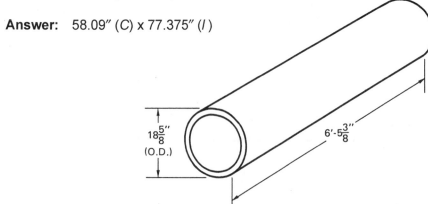

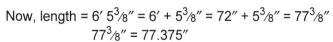

PRACTICAL PROBLEMS

1. Find the size of the 20-gauge sheet metal needed to construct this
    semicircular ventilation section. The average diameter is $19\frac{3}{16}$″.   _____

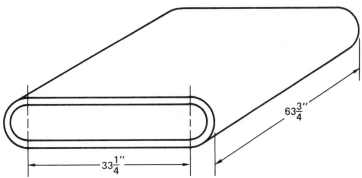

2. This hydraulic ram cylinder shown below is rolled from 1.5875 cm thick
    metal. Find the size of the plate steel needed to construct the cylinder.   _____

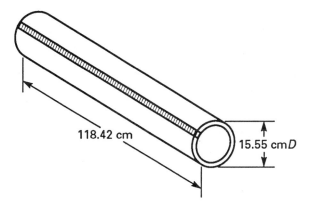

3.   This semicircular-sided tank is rolled from $^3/_{16}''$ plate. The average
     diameter is $19^{11}/_{16}''$.

_____

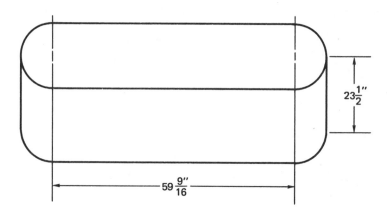

a.   Find the amount of $^3/_{16}''$ plate needed to roll the tank.   a. _____

b.   The bottom of the tank is cut from a rectangular piece of $^3/_{16}''$ plate.
     Find the width of the plate.   b. _____

c.   Find the length of plate needed for the bottom of the tank.   c. _____

4.   This electrode holding-tube has an average radius of 6.66 cm. Find the
     size of the single sheet of 22-gauge metal needed to construct 10 tubes.   _____

5.   A branch header is constructed as shown. Find the stretchout of the material needed to construct the two pieces of the branch header from $\frac{3}{16}$″ steel plate. Make no allowances for weld gaps or seams.

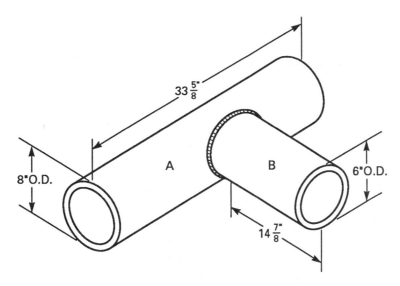

a.   Branch header A

b.   Branch header B

a. _____

b. _____

6.   The average diameter of this storage tank is 9 feet $6\frac{1}{2}$″. Find the number of $\frac{1}{8}$″ sheet metal pieces needed to complete the cylindrical side of this storage tank. The sheet size available is $\frac{1}{8}$″ x 48″ x 97.735″.

_____

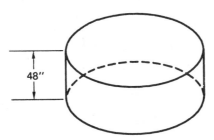

Unit 39 *ECONOMICAL LAYOUT OF RECTANGULAR PLATES*

BASIC PRINCIPLES

To find the most economical layout of a steel plate, make a sketch, laying out the length and width of the cut pieces both ways on the plate.

Example: How many pieces 3″ x 4″ may be cut from a plate 9″ x 17″?

Make a sketch:

	17″				2″ WASTE
3″	**3″**	**3″**	**3″**	**3″**	
1	2	3	4	5	
6	7	8	9	10	

4″ / 9″ / 4″ on the left side, 1″ WASTE at the bottom.

10 PIECES

	17″			1″ WASTE
4″	**4″**	**4″**	**4″**	
1	2	3	4	
5	6	7	8	
9	10	11	12	

9″ with 3″ / 3″ / 3″ rows on the left side.

12 PIECES

162

PRACTICAL PROBLEMS

Answers should show the maximum number of pieces that can be obtained. Disregard waste caused by the width of the cuts.

1.   A weld shop supplies 104 shaft blanks, each 4″ wide and 5″ long. How many can be cut from the piece of plate shown?

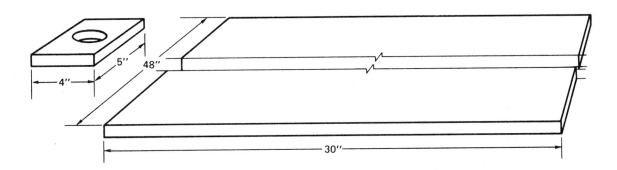

2.   How many 12.7 cm by 15.24 cm plates can be cut from this plate?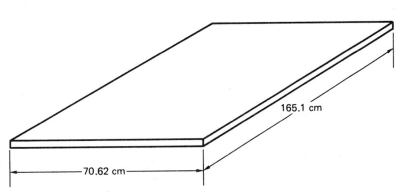

3. These pulley brackets are cut from a 1″ thick plate of steel that is $6\frac{1}{2}$″ by $48\frac{1}{2}$″.

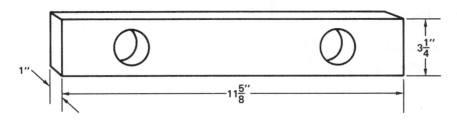

1″

$3\frac{1}{4}$″

$11\frac{5}{8}$″

a. How many of these brackets can be cut?   a. _____

b. What is the size of the material remaining after the brackets are cut?   b. _____

4. Column baseplates of the size shown are cut and drilled. The baseplates are cut from a steel plate with the dimensions of 84.750″ by 89.500″.

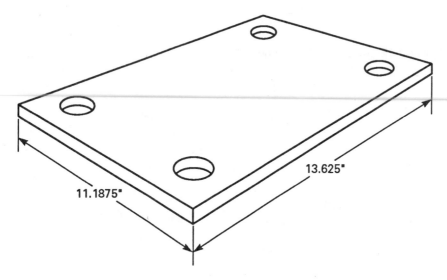

13.625″

11.1875″

a. How many baseplates can be cut?   a. _____

b. How much scrap material remains after the baseplates are cut?   b. _____

5.   These angle brackets are cut and welded to finish a construction job. How many of the brackets can be obtained from a steel plate that is 60″ wide by 90½″ long?   _____

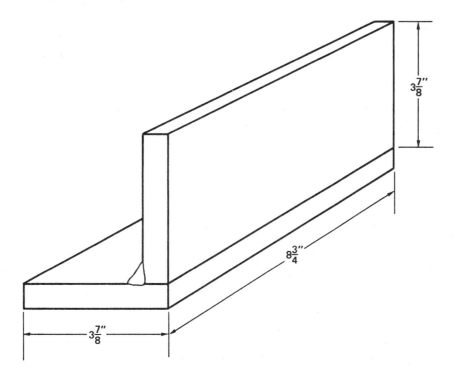

6.   Storage bin sides are 20.32 cm by 27.94 cm. Three sides are welded together for each bin. How many bins can be made from the plate of steel shown?   _____

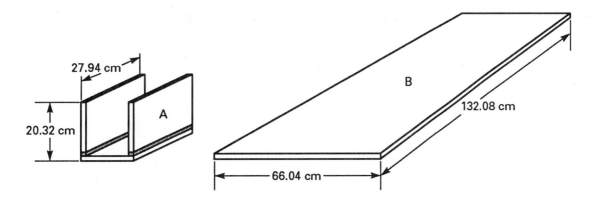

Find the maximum number of pieces of the given sizes that can be cut from the indicated sheets of steel.

	Dimensions of Piece	Sheet Size	Maximum Number of Pieces
7.	7 in × 8 in	36 in × 71 in	_____
8.	8 in × 11 in	48 in × 120 in	_____
9.	25.4 cm × 33.02 cm	152.4 cm × 304.8 cm	_____
10.	21 in × 27 in	3 ft 9 in × 5 ft	_____
11.	5.08 cm × 15.24 cm	33.02 cm × 137.16 cm	_____

Unit 40   ECONOMICAL LAYOUT OF ODD-SHAPED PLATES

BASIC PRINCIPLES

Review the principles of economical layout of odd-shaped plates, and apply them to these problems.

Answers should show the maximum number of pieces that can be obtained. Sketches of different arrangements may be helpful. Disregard waste caused by the width of the cuts.

Example: A piece of scrap metal of the shape shown below is cut into 10″ radius circles. How many circles can be cut from the material?

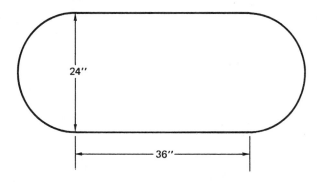

24″

36″

$r = 10$; $D = 20$ in

$24 \text{ in} \div \dfrac{20 \text{ in}}{1 \text{ circle}} = 1.2 \text{ circles or 1 circle}$

$36 \text{ in} + 12 \text{ in} + 12 \text{ in} = 60 \text{ in}$

$60 \text{ in} \div \dfrac{20 \text{ in}}{1 \text{ circle}} = 3 \text{ circles}$

$1 \times 3 = 3 \text{ circles}$

Note: Use this diagram for Problem 1.

PRACTICAL PROBLEMS

1. How many 13″ diameter circles can be cut from the scrap metal?   _____

2. How many pieces of sheet metal ¾″ wide and 60″ long can be cut from this sheet?   _____

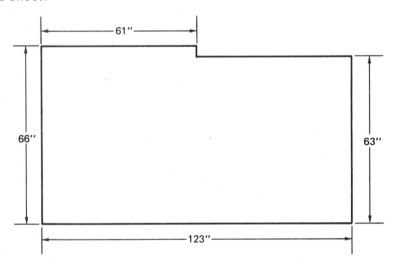

3. Fourteen sections of 2″ pipe of the length shown are used to construct a framework.

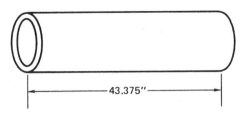

   a. How many standard 21′ lengths of pipe must be used to cut the 14 sections?   a. _____

   b. What percent of the pipe used is wasted after cutting? Round the answer to the nearest hundredth percent.   b. _____

4. Thirty-one column gussets are cut from a steel plate. Find the dimensions of the smallest square plate that can be used for the gussets.

_____

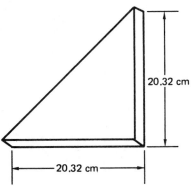

20.32 cm

20.32 cm

5. This circular blank is used to make sprocket drives. How many sprocket drive blanks can be cut from a plate of steel having the dimensions of 44″ x 44″?

_____

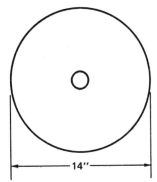

14″

6.   Gussets are cut from the material shown. How many gussets can be cut from one sheet?

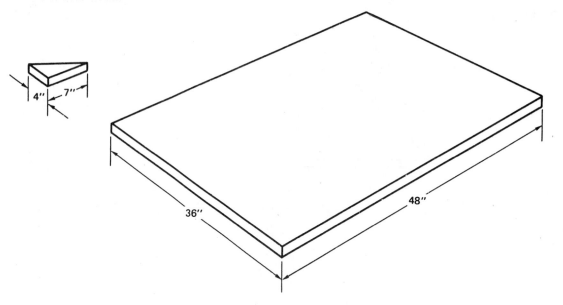

7.   How many 4″ x 3″ rectangular test plates can be cut from this piece of scrap?

_____

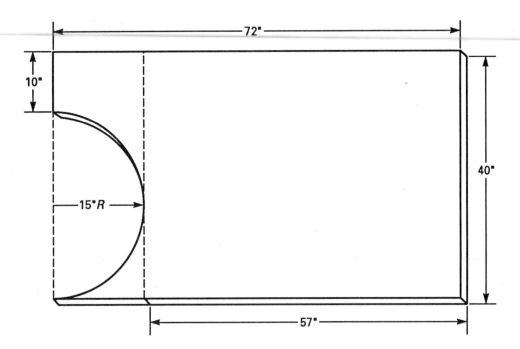

APPENDIX

SECTION I

DENOMINATE NUMBERS

Denominate numbers are numbers that include units of measurement. The units of measurement are arranged from the largest units at the left to the smallest units at the right.

For example: 6 yd 2 ft 4 in

All basic operations of arithmetic can be performed on denominate numbers.

I. EQUIVALENT MEASURES

Measurements that are equal can be expressed in different terms. For example, 12 in = 1 ft. If these equivalents are divided, the answer is 1.

$$\frac{1 \text{ ft}}{12 \text{ in}} = 1 \qquad\qquad \frac{12 \text{ in}}{1 \text{ ft}} = 1$$

To express one measurement as another equal measurement, multiply by the equivalent in the form of 1.

To express 6 inches in equivalent foot measurement, multiply 6 inches by one in the form of $\frac{1 \text{ ft}}{12 \text{ in}}$. In the numerator and denominator, divide by a common factor.

$$6 \text{ in} = \frac{\overset{1}{\cancel{6 \text{ in}}}}{1} \times \frac{1 \text{ ft}}{\underset{2}{\cancel{12 \text{ in}}}} = \frac{1}{2} \text{ ft or 0.5 ft.}$$

To express 4 feet in equivalent inch measurement, multiply 4 feet by one in the form of $\frac{12 \text{ in}}{1 \text{ ft}}$.

$$4 \text{ ft} = \frac{\overset{4}{\cancel{4 \text{ ft}}}}{1} \times \frac{12 \text{ in}}{\underset{1}{\cancel{1 \text{ ft}}}} = \frac{48 \text{ in}}{1} = 48 \text{ in}$$

Per means division, as with a fraction bar. For example, 50 miles per hour can be written $\frac{50 \text{ miles}}{1 \text{ hour}}$

II. BASIC OPERATIONS

A. ADDITION

SAMPLE: 2 yd 1 ft 5 in + 1 ft 8 in + 5 yd 2 ft

1.	Write the denominate numbers in a column with like units in the same column.	2 yd + 5 yd	1 ft 1 ft 2 ft	5 in 8 in

2. Add the denominate numbers in each column.

        7 yd   4 ft   13 in

3. Express the answer using the largest possible units.

        7 yd   = 7 yd

               4 ft   = 1 yd   1 ft

                     13 in =+   1 ft   1 in

        7 yd   4 ft   13 in = 8 yd   2 ft   1 in

B. SUBTRACTION

SAMPLE: 4 yd 3 ft 5 in - 2 yd 1 ft 7 in

1.	Write the denominate numbers in columns with like units in the same column.	4 yd − 2 yd	3 ft 1 ft	5 in 7 in

2. Starting at the right, examine each column to compare the numbers. If the bottom number is larger, exchange one unit from the column at the left for its equivalent. Combine like units.

        7 in is larger than 5 in

        3 ft = 2 ft 12 in

        12 in + 5 in = 17 in

3. Subtract the denominate numbers.

        4 yd   2 ft   17 in

    −   2 yd   1 ft   7 in

        2 yd   1 ft   10 in

4. Express the answer using the largest possible units.

        2 yd   1 ft   10 in

C. MULTIPLICATION

– *By a constant*
SAMPLE: 1 hr 24 min X 3

1. Multiply the denominate number
   by the constant.

$$
\begin{array}{ll}
1\ hr & 24\ min \\
& x\ \ 3 \\
\hline
3\ hr & 72\ min
\end{array}
$$

2. Express the answer using the
   largest possible units.

$$
\begin{array}{ll}
3\ hr & \qquad = 3\ hr \\
\quad\ \ 72\ min & \qquad = 1\ hr \quad 12\ min \\
\hline
3\ hr \quad 72\ min & \qquad = 4\ hr \quad 12\ min
\end{array}
$$

– *By a denominate number expressing linear measurement*
SAMPLE: 9 ft 6 in X 10 ft

1. Express all denominate numbers
   in the same unit.

$$9\ ft\ 6\ in = \quad 9\ \tfrac{1}{2}\ ft$$

2. Multiply the denominate numbers.
   (This includes the units of measure,
   such as ft X ft = sq ft).

$$9\ \tfrac{1}{2}\ ft\ X\ 10\ ft =$$

$$\tfrac{19}{2}\ ft\ X\ 10\ ft =$$

95 sq ft

– *By a denominate number expressing square measurement*
SAMPLE: 3 ft X 6 sq ft

1. Multiply the denominate numbers.
   (This includes the units of measure,
   such as ft X ft = sq ft and sq ft
   X ft = cu ft).

3 ft X 6 sq ft = 18 cu ft

– *By a denominate number expressing rate*
SAMPLE: 50 miles per hour X 3 hours

1. Express the rate as a fraction using
   the fraction bar for *per.*

$$\frac{50\ miles}{1\ hour}\ X\ \frac{3\ hours}{1}\ =$$

2. Divide the numerator and denominator
   by any common factors, including
   unitsofmeasure.

$$\frac{50\ miles}{\cancel{1\ hour}}\ X\ \frac{\cancel{3\ hours}^{\ 3}}{1}\ =$$

3. Multiply   numerators.
    Multiply   denominators.

$$\frac{150 \text{ miles}}{1} =$$

4. Express the answer in the remaining
    unit.

150 miles

D. DIVISION

– By a constant
SAMPLE: 8 gal 3 qt ÷ 5

1. Express all denominate numbers
    in the same unit.

8 gal 3 qt = 35 qt

2. Divide the denominate number
    by the constant.

35 qt ÷ 5 = 7 qt

3. Express the answer using the
    largest possible units.

7 qt = 1 gal 3 qt

– By a denominate number expressing linear measurement
SAMPLE: 11 ft 4 in ÷ 8 in

1. Express all denominate numbers
    in the same unit.

11 ft 4 in = 136 in

2. Divide the denominate numbers
    by a common factor. (This includes
    the units of measure, such as inches
    ÷ inches = 1.)

136 in ÷ 8 in =

$$\frac{\overset{17}{\cancel{136 \text{ in}}}}{\underset{1}{\cancel{8 \text{ in}}}} = \frac{17}{1} = 17$$

– By a linear measure with a square measurement as the dividend
SAMPLE: 20 sq ft ÷ 4 ft

1. Divide the denominate numbers.
    (This includes the units of measure,
    such as sq ft ÷ ft = ft)

20 sq ft ÷ 4 ft

$$\frac{\overset{5 \text{ ft}}{\cancel{20 \text{ sq ft}}}}{\cancel{4 \text{ ft}}} = \frac{5 \text{ ft}}{1}$$

2. Express the answer in the remain-
    ing unit.

5 ft

– By denominate numbers used to find rate

SAMPLE: 200 mi ÷ 10 gal

1. Divide the denominate numbers.

$$\frac{\cancel{200}^{\;20} \text{ mi}}{\cancel{10}^{\;1} \text{ gal}} = \frac{20 \text{ mi}}{1 \text{ gal}}$$

2. Express the units with the fraction bar meaning *per*.

$$\frac{20 \text{ mi}}{1 \text{ gal}} = 20 \text{ miles per gallon}$$

Note: Alternate methods of performing the basic operations will produce the same result. The choice of method is determined by the individual.

SECTION II

EQUIVALENTS

ENGLISH RELATIONSHIPS

ENGLISH LENGTH MEASURE

1 foot (ft.)	=	12 inches (in.)
1 yard (yd.)	=	3 feet (ft.)
1 mile (mi.)	=	1,760 yards (yd.)
1 mile (mi.)	=	5,280 feet (ft.)

ENGLISH AREA MEASURE

1 square yard (sq. yd.)	=	9 square feet (sq. ft.)
1 square foot (sq. ft.)	=	144 square inches (sq. in.)
1 square mile (sq. mi.)	=	640 acres
1 acre	=	43,560 square feet (sq. ft.)

ENGLISH VOLUME MEASURE FOR SOLIDS

1 cubic yard (cu. yd.)	=	27 cubic feet (cu. ft.)
1 cubic foot (cu. ft.)	=	1,728 cubic inches (cu. in.)

ENGLISH VOLUME MEASURE FOR FLUIDS

1 quart (qt.)	=	2 pints (pt.)
1 gallon (gal.)	=	4 quarts (qt.)

ENGLISH VOLUME MEASURE EQUIVALENTS

1 gallon (gal.)	=	0.133681 cubic foot (cu. ft.)
1 gallon (gal.)	=	231 cubic inches (cu. in.)

DECIMAL AND METRIC EQUIVALENTS OF FRACTIONS OF AN INCH

FRACTION	1/32nds	1/64ths	DECIMAL	MILLIMETERS
		1	0.015625	0.3968
	1	2	0.03125	0.7937
		3	0.046875	1.1906
1/16	2	4	0.0625	1.5875
		5	0.078125	1.9843
	3	6	0.09375	1.3812
		7	0.109375	2.7780
1/8	4	8	0.125	3.1749
		9	0.140625	3.5718
	5	10	0.15625	3.9686
		11	0.171875	4.3655
3/16	6	12	0.1875	4.7624
		13	0.203125	5.1592
	7	14	0.21875	5.5561
		15	0.234375	5.9530
1/4	8	16	0.250	6.3498
		17	0.265625	6.7467
	9	18	0.281 25	7.1436
		19	0.296875	7.5404
5/16	10	20	0.3125	7.9373
		21	0.328125	8.3341
	11	22	0.343 75	8.7310
		23	0.359375	9.1279
3/8	12	24	0.375	9.5240
		25	0.390625	9.9216
	13	26	0.406 25	10.3185
		27	0.421875	10.7154
7/16	14	28	0.4375	11.1122
		29	0.453125	11.5091
	15	30	0.46875	11.9060
		31	0.484375	12.3029
1/2	16	32	0.500	12.6997
		33	0.515625	13.0966
	17	34	0.53125	13.4934
		35	0.546875	13.8903
9/16	18	36	0.5625	14.2872

DECIMAL AND METRIC EQUIVALENTS OF FRACTIONS OF AN INCH (cont.)

FRACTION	1/32nds	1/64ths	DECIMAL	MILLIMETERS
		37	0.578125	14.6841
	19	38	0.59375	15.0809
		39	0.609375	15.4778
5/8	20		0.625	15.8747
		41	0.640625	16.2715
	21	42	0.65625	16.6684
		43	0.671875	17.0653
11/16	22	44	0.6875	17.4621
		45	0.703125	17.8590
	23	46	0.71875	18.2559
		47	0.734375	18.6527
3/4	24	48	0.750	19.0496
		49	0.765625	19.4465
	25	50	0.78125	19.8433
		51	0.796875	20.2402
13/16	26	52	0.8125	20.6371
		53	0.828125	21.0339
	27	54	0.84375	21.4308
		55	0.859375	21.8277
7/8	28	56	0.875	22.2245
		57	0.890625	22.6214
	29	58	0.90625	23.0183
		59	0.921875	23.4151
5/16	30	60	0.9375	23.8120
		61	0.953125	24.2089
	31	62	0.96875	24.6057
		63	0.984375	25.0026
1	32	64	1.000	25.3995

SI METRICS STYLE GUIDE

SI metrics is derived from the French name Système International d'Unites. The metric unit names are already in accepted practice. SI metrics attempts to standardize the names and usages so that students of metrics will have a universal knowledge of the application of terms, symbols, and units.

The English system of mathematics (used in the United States) has always had many units in its weights and measures tables that were not applied to everyday use. For example, the pole, perch, furlong, peck, and scruple are not used often. These measurements, however, are used to form other measurements and it has been necessary to include the measurements in the tables. Including these measurements aids in the understanding of the orderly sequence of measurements greater or smaller than the less frequently used units.

The metric system also has units that are not used in everyday application. Only by learning the lesser-used units is it possible to understand the order of the metric system. SI metrics, however, places an emphasis on the most frequently used units.

In using the metric system and writing its symbols, certain guidelines are followed. For the students' reference, some of the guidelines are listed.

1. In using the symbols for metric units, the first letter is capitalized only if it is derived from the name of a person.

SAMPLE:	UNIT	SYMBOL	UNIT	SYMBOL
	meter	m	Newton	N (named after Sir Isaac Newton)
	gram	g	degree Celsius	°C (named after Anders Celsius)
EXCEPTION:	The symbol for liter is L. This is used to distinguish it from the number one (1).			

2. Prefixes are written with lowercase letters.

SAMPLE:	PREFIX	UNIT	SYMBOL
	centi	meter	cm
	milli	gram	mg
EXCEPTIONS:	PREFIX	UNIT	SYMBOL
	tera	meter	Tm (used to distinguish it from the metric ton,t)
	giga	meter	Gm (used to distinguish it from gram, g)
	mega	gram	Mg (used to distinguish it from milli, m)

3. Periods are not used in the symbols. Symbols for units are the same in the singular and the plural (no "s" is added to indicate a plural).

    SAMPLE: 1 mm *not* 1 mm.   3 mm *not* 3 mms

4. When referring to a unit of measurement, symbols are not used. The symbol is used only when a number is associated with it.

SAMPLE:	The length of the room is expressed in meters.	*not*	The length of the room is expressed in m. (*The length of the room is 25 m* is correct.)

5. When writing measurements that are less than one, a zero is written before the decimal point.

    SAMPLE: 0.25 m *not* .25 m

6. Separate the digits in groups of three, using commas to the left of the decimal point but not to the right.

    SAMPLE: 5,179,232 mm *not* 5 179 232 mm   0.56623 mg *not* 0.566 23 mg
    1,346.098,7 L *not* 1 346.098 7 L

    A space is left between the digits and the unit of measure.

    SAMPLE: 5,179,232 mm *not* 5,179,232mm

7. Symbols for area measure and volume measure are written with exponents.

    SAMPLE: 3 cm^2 *not* 3 sq cm   4 km^3 *not* 4 cu km

8. Metric words with prefixes are accented on the first syllable. In particular, kilometer is pronounced "kill′-o-meter." This avoids confusion with words for measuring devices that are generally accented on the second syllable, such as thermometer (ther-mom′-e-ter).

METRIC RELATIONSHIPS

The base units in SI metrics include the meter and the gram. Other units of measure are related to these units. The relationship between the units is based on powers of ten and uses these prefixes:

kilo (1,000)   hecto (100)   deka (10)   deci (0.1)   centi (0.01)   milli (0.001)

These tables show the most frequently used units with an asterisk (*).

METRIC LENGTH

10 millimeters (mm)*	=	1 centimeter (cm)*
10 centimeters (cm)	=	1 decimeter (dm)
10 decimeters (dm)	=	1 meter (m)*
10 meters (m)	=	1 dekameter (dam)
10 dekameters (dam)	=	1 hectometer (hm)
10 hectometers (hm)	=	1 kilometer (km)*

◆ To express a metric length unit as a smaller metric length unit, multiply by a positive power of ten such as 10, 100, 1,000, 10,000 etc.

◆ To express a metric length unit as a larger metric length unit, multiply by a negative power of ten such as 0.1, 0.001, 0.001, 0.0001, etc.

METRIC AREA MEASURE

100 square millimeters (mm^2)	=	1 square centimeter (cm^2)*
100 square centimeters (cm^2)	=	1 sqaure decimeter (dm^2)
100 square decimeters (cm^2)	=	1 square meter (m^2)*
100 square meters (m^2)	=	1 square dekameter (dam^2)
100 square dekameters (dam^2)	=	1 square hectometer (hm^2)*
100 square hectometers (hm^2)	=	1 square kilometer (km^2)

◆ To express a metric area unit as a smaller metric area unit, multiply by 100, 10,000, 1,000,000, etc.

◆ To express a metric area unit as a larger metric area unit, multiply by 0.01, 0.0001, 0.000001, etc.

METRIC VOLUME MEASURE FOR SOLIDS

1,000 cubic millimeters (mm^3)	=	1 cubic centimeter (cm)*
1,000 cubic centimeters (cm^3)	=	1 cubic decimeter (dm^3)*
1,000 cubic decimeters (dm^3)	=	1 cubic meter (m^3)*
1,000 cubic meters (m^3)	=	1 cubic dekameter (dam^3)
1,000 cubic dekameters (dam^3)	=	1 cubic hectometer (hm^3)
1,000 cubic hectometers (hm^3)	=	1 cubic kilometer (km^3)

◆ To express a metric volume unit for solids as a smaller metric volume unit for solids, multiply by 1,000, 1,000,000, 1,000,000,000, etc.

◆ To express a metric volume unit for solids as a larger metric volume unit for solids, multiply by 0.001, 0.000001, 0.000000001, etc.

METRIC VOLUME MEASURE FOR FLUIDS

10 milliliters (mL)*	=	1 centiliter (cL)
10 centiliters (cL)	=	1 deciliter (dL)
10 deciliters (dL)	=	1 liter (L)*
10 liters (L)	=	1 dekaliter (daL)
10 dekaliters (daL)	=	1 hectoliter (hL)
10 hectoliters (hL)	=	1 kiloliter (kL)

◆ To express a metric volume unit for fluids as a smaller metric volume unit for fluids, multiply by 10, 100, 1,000, 10,000, etc.

◆ To express a metric volume unit for fluids as a larger metric volume unit for fluids, multiply by 0.1, 0.01, 0.001, 0.0001, etc.

METRIC VOLUME MEASURE EQUIVALENTS

1 cubic decimeter (dm^3)	=	1 liter (L)
1,000 cubic centimeters (cm^3)	=	1 liter (L)
1 cubic centimeter (cm^3)	=	1 milliliter (mL)

METRIC MASS MEASURE

10 milligrams (mg)*	=	1 centigram (cg)
10 centigrams (cg)	=	1 decigram (dg)
10 decigrams (dg)	=	1 gram (g)*
10 grams (g)	=	1 dekagram (dag)
10 dekagrams (dag)	=	1 hectogram (hg)
10 hectograms (hg)	=	1 kilogram (kg)*
1,000 kilograms (kg)	=	1 megagram (Mg)*

◆ To express a metric mass unit as a smaller metric mass unit, multiply by 10, 100, 1,000, 10,000, etc.
◆ To express a metric mass unit as a larger metric mass unit, multiply by 0.1, 0.01, 0.001, 0.0001, etc.

Metric measurements are expressed in decimal parts of a whole number. For example, one-half millimeter is written as 0.5 mm.

In calculating with the metric system, all measurements are expressed using the same prefixes. If answers are needed in millimeters, all parts of the problem should be expressed in millimeters before the final solution is attempted. Diagrams that have dimensions in different prefixes must first be expressed using the same unit.

ENGLISH-METRIC EQUIVALENTS

LENGTH MEASURE

1 inch (in)	=	25.4 millimeters (mm)
1 inch (in)	=	2.54 centimeters (cm)
1 foot (ft)	=	0.30,8 meter (m)
1 yard (yd)	=	0.9144 meter (m)
1 mile (mi)	≈	1.609 kilometers (km)
1 millimeter (mm)	≈	0.03937 inch (in)
1 centimeter (cm)	≈	0.39370 inch (in)
1 meter (m)	≈	3.28084 feet (ft)
1 meter (m)	≈	1.09361 yards (yd)
1 kilometer (km)	≈	0.62137 mile (mi)

AREA MEASURE

1 square inch (sq in)	=	645.16 square millimeters (mm^2)
1 square inch (sq in)	=	6.4516 square centimeters (cm^2)
1 square foot (sq ft)	≈	0.092903 square meter (m^2)
1 square yard (sq yd)	≈	0.836127 square meter (m^2)
1 square millimeter (mm^2)	≈	0.001550 square inch (sq in)
1 square centimeter (cm^2)	≈	0.15500 square inch (sq in)
1 square meter (m^2)	≈	10.763910 square feet (sq ft)
1 square meter (m^2)	≈	1.19599 square yards (sq yd)

VOLUME MEASURE FOR SOLIDS

1 cubic inch (cu in)	=	16.387064 cubic centimeters (cm^3)
1 cubic foot (cu ft)	≈	0.028317 cubic meter (m^3)
1 cubic yard (cu yd)	≈	0.764555 cubic meter (m^3)
1 cubic centimeter (cm^3)	≈	0.061024 cubic inch (cu in)
1 cubic meter (m^3)	≈	35.314667 cubic feet (cu ft)
1 cubic meter (m^3)	≈	1.307951 cubic yards (cu yd)

VOLUME MEASURE FOR FLUIDS

1 gallon (gal)	≈	3,785.411 cubic centimeters (cm^3)
1 gallon (gal)	≈	3.785411 litres (L)
1 quart (qt)	≈	0.946353 liter (L)
1 ounce (oz)	≈	29.573530 cubic centimeters (cm^3)
1 cubic centimeter (cm^3)	≈	0.000264 gallon (gal)
1 liter (L)	≈	0.264172 gallon (gal)
1 litre (L)	≈	1.056688 quarts (qt)
1 cubic centimeter (cm^3)	≈	0.033814 ounce (oz)

MASS MEASURE

1 pound (lb)	≈	0.453592 kilogram (kg)
1 pound (lb)	≈	453.59237 grams (g)
1 ounce(oz)	≈	28.349523 grams (g)
1 ounce(oz)	≈	0.028350 kilogram (kg)
1 kilogram (kg)	≈	2.204623 pounds (lb)
1 gram (g)	≈	0.002205 pound (lb)
1 kilogram (kg)	≈	35.273962 ounces (oz)
1 gram (g)	≈	0.035274 ounce (oz)

SECTION III

FORMULAS

Perimeter	P = perimeter	**Area**	A = area
Square $P = 4s$	P = perimeter s = side	Square $A = s \times s$	s = length of side
Rectangle $P = 2l + 2w$	P = perimeter l = length w = width	Rectangle $A = lw$	l = length w = width
		Triangle $A = \frac{1}{2}bh$	h = height b = base
Circle $C = \pi D$ or $C = 2\pi r$	C = circumference π = 3.14 D = diameter r = radius = $\frac{1}{2}D$	Trapezoid $A = \frac{1}{2}(B + b)h$	B = length of large side b = length of small side h = height
		Circle $A = \pi r^2$ or $A = \frac{\pi}{4}D^2$	π = 3.14 r = radius D = diameter
Semicircular-sided figure $P = \pi D + 2l$	P = perimeter π = 3.14 D = diameter l = length	Semicircular-sided figure $A = \pi r^2 + Dl$	π = 3.14 D = diameter l = length r = radius = $\frac{1}{2}D$

Volume	V = Volume	**Stretchouts**	
Cube $V = s^3$	s = side	Square pipe LS $= 4s$ WS $= h$	s = side h = height
Rectangular solid $V = lwh$	l = length w = width h = height	Rectangular pipe $LS = 2l + 2w$ $WS = h$	LS = length of stretchout WS = width of stretchout l = length w = width
Cylinder $V = \pi r^2 l$	π = 3.14 r = radius l = length	Circular Pipe LS $= \pi D$ WS $= h$	LS = length of stretchout WS = width of stretchout D = diameter h = height π = 3.14
Semicircular-sided solid $V = (\pi r^2 + Dl)h$	π = 3.14 r = radius D = diameter l = distance between centers of semicircle h = height	Semicircle tanks LS $= \pi D + 2l$ WS $= h$	π = 3.14 D = diameter l = distance between centers of semicircles h = height

Bend Allowance

(Approximate allowances)

For inside dimensions: T = thickness
Length of stretchout = L = long leg
  L_1 = short leg

$$L + L_1 + \frac{T}{2}$$

For outside dimensions:
Length of stretchout = $L + L_1 - \frac{T}{2}$

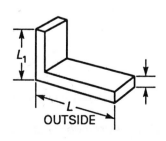

OUTSIDE

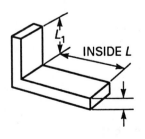

INSIDE L

GLOSSARY

Angle — A piece of steel rolled into the shape of the letter L; used for forming the joints in girders, boilers, and so forth.

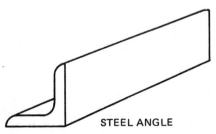

STEEL ANGLE

Channel — A flanged steel beam whose section forms three sides of a parallelogram; used for structural purposes.

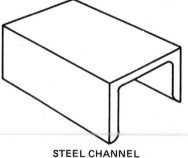

STEEL CHANNEL

Cold-rolled steel — Steel that is rolled from a cold bar of steel. Cold rolling produces steel to closer tolerances than hot rolling at the mill. Can be easily identified by its clean appearance and lack of "mill scale."

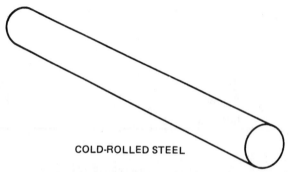

COLD-ROLLED STEEL

Flame cutting — Cutting of steel performed by an oxygen-fuel gas torch flame having an oxygen jet.

Flat bar — Flats are furnished in different widths and generally in either 18- or 20-foot lengths. Some companies designate the material as "flat" until it exceeds 6 inches in width, and others furnish it in 8-inch or 10-inch widths.

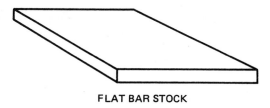

FLAT BAR STOCK

I beam — A beam that has a cross section shaped like the letter I.

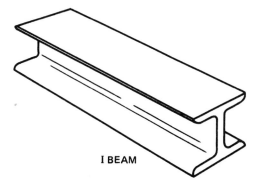

I BEAM

Mild steel — As a rule, a steel with a carbon range between 0.05 percent and 0.30 percent is called low-carbon steel, or mild steel. Mild steel is most commonly used for welding and appears in most welded structures.

Pipe — A long tube or hollow body used for the conveyance of all types of fluids and gases. Pipe and tubing are standard structural shapes used in welding.

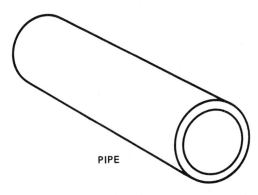

PIPE

Plate — Steel that has been rolled in thicknesses of $\frac{1}{8}$ inch or more is generally designated as plate.

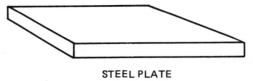

STEEL PLATE

Round bar — Round bar and other steel shapes that are ordered in cold-rolled steel are exact in size. If ordered in hot-rolled, the bars are generally a small amount oversized, but cheaper in price. Cold-rolled steel has a shiny, nickel-like appearance; hot-rolled steel is dark and has some "mill scale" on it.

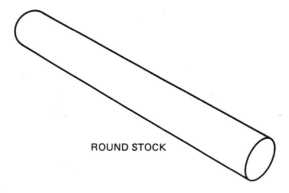

ROUND STOCK

Sheet metal — Metal that has been rolled in the steel mill to a thickness of less than $\frac{1}{8}$ inch.

RECTANGULAR SHEET STEEL

Square tubing — A structural shape used for many purposes. Although it is generally called square, the sides may vary in size so that it appears as a rectangular hollow structure.

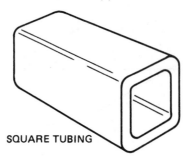

SQUARE TUBING

Wide-flange beam — Formed in the shape of an I, wide-flange has parallel flanges.

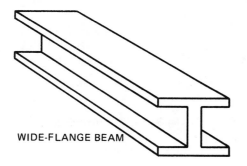

WIDE-FLANGE BEAM

ANSWERS TO ODD-NUMBERED PROBLEMS

SECTION 1 WHOLE NUMBERS

UNIT 1 ADDITION OF WHOLE NUMBERS

1. 1,502 ft

3. 11,323 lb

UNIT 2 SUBTRACTION OF WHOLE NUMBERS

1. $61\frac{7}{8}$ in.

3. $5\frac{3}{4}$ in

UNIT 3 MULTIPLICATION OF WHOLE NUMBERS

1. a. 5,280 in
   b. 8,580 in

1. c. 2,640 lb

3. 9,882 in

UNIT 4 DIVISION OF WHOLE NUMBERS

1. 7 pcs
3. $\frac{25}{8}$ in = $3\frac{1}{8}$ in

5. a. 900 gal
   b. 383 lb

5. c. $1,036

SECTION 2 COMMON FRACTIONS

UNIT 5 INTRODUCTION TO THE STEEL TAPE AND COMMON FRACTIONS

1. a. $\frac{5}{16}$
   b. $\frac{8}{16}$ or $\frac{1}{2}$
   c. $\frac{13}{16}$

3. $\frac{4}{8}$ or $\frac{1}{2}$ in
5. E. $\frac{1}{8}$ in
   F. $\frac{3}{8}$ in

5. G. $\frac{5}{8}$ in
   H. $\frac{7}{8}$ in

UNIT 6 ADDITION OF COMMON FRACTIONS

1. $7\frac{3}{16}$ in
3. $12\frac{11}{16}$ in

5. a. $8\frac{1}{16}$ in
   b. $4\frac{1}{8}$ in

7. $1\frac{7}{8}$ in
9. a. $2\frac{13}{16}$ in
   b. $4\frac{13}{16}$ in.

UNIT 7 SUBTRACTION OF COMMON FRACTIONS

1. $8\frac{3}{16}$ in
3. $7\frac{3}{16}$ in

5. $5\frac{1}{16}$ in

7. $3\frac{9}{16}$ in

UNIT 8 MULTIPLICATION OF COMMON FRACTIONS

1. a. $81\frac{1}{2}$ in
    b. 1,630 in

3. $131\frac{11}{16}$ in
5. $21\frac{7}{8}$ in

7. $111\frac{3}{8}$ rods

UNIT 9 DIVISION OF COMMON FRACTIONS

1. 3 pcs
3. $3\frac{5}{16}$ in

5. 48 strips

7. 32 pcs

UNIT 10 COMBINED OPERATIONS WITH COMMON FRACTIONS

1. $6\frac{1}{2}$ in

3. 5 in

5. $1\frac{1}{4}$ in

SECTION 3 DECIMAL FRACTIONS

UNIT 11 INTRODUCTION TO DECIMAL FRACTIONS

1. a. 0.3
    b. 0.6

1. c. 0.9

3. 26.048 gal

UNIT 12 ADDITION AND SUBTRACTION OF DECIMAL FRACTIONS

1. 5.1875 in

3. 1.25 in

5. 36.75 lb

UNIT 13 MULTIPLICATION OF DECIMAL FRACTIONS

1. 308.75 in
3. 60.5625 cu ft

5. 36.751 lb
7. 4.875 lb

9. 18.419 in

UNIT 14 DIVISION OF DECIMAL FRACTIONS

1. 23 sockets
3. 484 base plates

5. 0.6571 in
7. 26 pcs

9. 4.37 in

UNIT 15 DECIMAL AND COMMON FRACTION EQUIVALENTS

1. a. 4.0625 in
    b. 6.25 in
    c. 0.375 in
3. a. $10\frac{3}{16}$ in
    b. $2\frac{3}{4}$ in
    c. $3\frac{1}{4}$ in
    d. $\frac{3}{8}$ in
5. a. $11\frac{1}{8}$ in
    b. $2\frac{1}{64}$ in

7. a. $\frac{1}{16}$ in
    b. $\frac{3}{4}$ in
    c. $\frac{3}{16}$ in
    d. $\frac{1}{4}$ in
    e. $1\frac{7}{8}$ in
    f. $1\frac{3}{8}$ in
    g. $1\frac{3}{4}$ in

9. a. $\frac{11}{64}$
    b. $\frac{5}{16}$
    c. $\frac{7}{8}$
    d. $\frac{9}{16}$
    e. $\frac{15}{16}$
    f. $\frac{33}{64}$

UNIT 16 TOLERANCES

1. A = 9′ 8¾ in
    B = 9′ 8⁵⁄₁₆ in
3. A = 11′ 9⅞ in
    B = 11′ 8¹³⁄₁₆ in

5. A = 18′ 4⁹⁄₁₆ in
    B = 18′ 4⁷⁄₁₆ in

7. A = 10.160 cm
    B = 10.160 cm

UNIT 17 COMBINED OPERATIONS WITH DECIMAL FRACTIONS

1. 6.3125 in
3. 302.25 in

5. 9 supports
7. 4.33 in

9. 0.39 cm

SECTION 4 AVERAGES, PERCENTS, AND PERCENTAGES

UNIT 18 AVERAGES

1. 10.729 in
3. 40 electrodes

5. 1.297 in

7. 35 lb

UNIT 19 PERCENTS AND PERCENTAGES

1. a. 0.01
   b. 0.05
   c. 0.08
   d. 0.60

1. e. 0.2325
   f. 1.25
   g. 2.20

3. 361.6125 in^2
5. a. 2,320 welds
   b. 95,120 welds
   c. 18,560 welds

SECTION 5 DIRECT MEASURE

UNIT 20 DIRECT MEASURE INSTRUMENTS

1. A = ⁴⁄₈ in (½ in)
3. E = ⅛ in
    F = ⅜ in

3. G = ⅝ in
    H = ⅞ in

5. Instructor to measure
    lines.

UNIT 21 METRIC LENGTH MEASURE

1. 2.2 cm
3. a. 2184.4 mm
   b. 25.4 mm
   c. 2.2352
5. 18 cm
7. 218 cm

9. A = 5 mm
    B = 8 mm
    C = 16 mm
    D = 23 mm
    E = 37 mm
    F = 53 mm
    G = 77 mm
    H = 93 mm

11. a. 16.51 cm
    b. 1,000 mm
    c. 60.96 mm
13. a. 73 cm
    b. 0.73 m

UNIT 22 EQUIVALENT UNITS

1. a. 12.7 cm
   b. 600 cm

3. $1\frac{1}{2}$ ft

5. $56\frac{3}{4}$ in

7. $1'\ 6\frac{5}{8}$ in

9. a. $3'\ 2\frac{3}{16}$ in
   b. $1'\ 4\frac{3}{8}$ in

UNIT 23 ENGLISH–METRIC EQUIVALENT UNITS

1. 2.500 m

3. a. 70.866 in
   b. 1.284 in

5. a. 47.466 cm
   b. 1.588 cm
   c. 17.939 cm

7. a. 1.5875 mm
   b. 3.175 mm
   c. 4.7625 mm
   d. 6.35 mm

7. e. 9.525 mm
   f. 12.7 mm
   g. 19.05 mm
   h. 17.463 mm

UNIT 24 COMBINED OPERATIONS WITH EQUIVALENT UNITS

1. 14.000 in

3. 320.675 mm

5. a. 144.463 cm
   b. 103.188 cm

7. a. 5,410.2 cm
   b. 3,214.2 cm
   c. 6,419.85 cm

UNIT 25 PERIMETER OF SQUARES AND RECTANGLES

1. 7 in

3. 774.7 mm

5. 97.536 cm

7. $10\frac{1}{2}$ in

UNIT 26 PERIMETER OF CIRCLES AND SEMICIRCULAR-SIDED FIGURES

1. 204.1 in

3. 117.96 in

SECTION 6 ANGULAR MEASURE

UNIT 27 ANGULAR MEASURE

1. 112° 14′ 8″

3. 271° 2′ 19″

5. 45° 22′ 32″

7. 67° 12′

9. 137° 23′ 10″

11. 270°

13. 225°

15. $\frac{1}{9}$

17. 0°

UNIT 28 PROTRACTORS

1. Instructor to measure angles.

SECTION 7 COMPUTED MEASURE

UNIT 29 AREA OF SOUARES AND RECTANGLES

1. $40^{41}\!/_{64}$ sq in
3. $162^{9}\!/_{16}$ sq in

5. 2,466,912.012 cm^2
7. C

9. 482.22 sq ft
11. 1 pc

UNIT 30 AREA OF TRIANGLES AND TRAPEZOIDS

1. 32 sq in
3. 72 sq in

5. 1,277.4168 cm^2
7. 75 sq in

9. 29 sq in
11. 562.5 sq ft

UNIT 31 AREA OF CIRCULAR FIGURES

1. 1,017.36 sq in
3. a. 6,723 sq in

3. b. 3,997.665 sq in
   c. 2,275.335 sq in

5. 1,160.96 sq in

UNIT 32 VOLUME OF CUBES AND RECTANGULAR SOLIDS

1. 1,728 cu in
3. 2.949 cu ft

5. 2,970.459 cu in
7. 294 cu in

9. 960 cu in
11. 4,960.881 cu in

UNIT 33 VOLUME OF CYLINDRICAL SOLIDS

1. 4,710 cu in
3. 3,102.858 cu in

5. 12.56 cu ft
7. 1.57 cu ft

9. 19,216.8 cu in
11. No

UNIT 34 VOLUME OF RECTANGULAR CONTAINERS

1. 6.381 gal
3. 134.649 gal
5. 12,167 cu in

7. 1,116.771 cu in
9. No
11. a. 691,200 cu in
    b. 400 cu ft

13. 52.0 in
15. 166.347 L

UNIT 35 VOLUME OF CYLINDRICAL AND COMPLEX CONTAINERS

1. a. 5,887.5 cu in
   b. 6,280 cu in
   c. 7,143.5 cu in

3. 864.010 gal
5. 0.533 cu m

7. 1,109 gal
9. 174.881 gal

UNIT 36 MASS (WEIGHT) MEASURE

1. 280.4100 lb
3. 105.8117 kg

5. 128.7806 lb
7. 133.9538 lb

9. 19.003 lb

SECTION 8 PIECES AND LENGTHS

UNIT 37 STRETCHOUTS OF SQUARE AND RECTANGULAR SHAPES

1. 169.2 cm x 317.5 cm
3. $75\frac{3}{4}$ in x 63 in

5. 60 bins

7. 113.3375 cm x 47.94 cm

UNIT 38 STRETCHOUTS OF CIRCULAR AND SEMICIRCULAR SHAPES

1. 126.749 in x 63.75 in
3. a. 180.994 in x 23.5 in

3. b. $19\frac{7}{8}$ in
   c. $79\frac{7}{16}$ in

5. a. 24.531 in
   b. 18.251 in

UNIT 39 ECONOMICAL LAYOUT OF RECTANGULAR PLATES

1. 72 blanks
3. a. 8 brackets
   b. $6\frac{1}{2}$ in x 2 in

5. 75 brackets
7. 40 pcs

9. 54 pcs
11. 54 pcs

UNIT 40 ECONOMICAL LAYOUT OF ODD-SHAPED PLATES

1. 6 circles

3. a. 3
   b. 19.68%

5. 9 sprocket blanks
7. 15 test plates